RADIOACTIVE WASTE FROM NUCLEAR POWER PLANTS

RADIOACTIVE WASTE FROM NUCLEAR POWER PLANTS

THOMAS B. JOHANSSON AND PETER STEEN

University of California Press
Berkeley   Los Angeles   London

University of California Press

Berkeley and Los Angeles, California

University of California Press, Ltd.

London, England

Library of Congress Cataloging in Publication Data

Johansson, Thomas B   1943–
   Radioactive waste from nuclear power plants.

   Translation of Kärnkraftens radioaktiva avfall—inför ringhals 3-beslutet.
   Includes index.
     1. Atomic power-plants—Sweden—Waste disposal.   2. Reactor fuel
reprocessing—Sweden—Waste disposal.   3. Atomic power-plants—Law and
legislation—Sweden.   4. Radioactive waste disposal—Law and legislation—
Sweden. I. Steen, Peter, 1943–   joint author. II. Title.
TD899.A8J6313   1981   363.7′28   80-6052
ISBN 0-520-04199-2

1 2 3 4 5 6 7 8 9

CONTENTS

Preface

In April 1977 the Swedish Parliament passed the Nuclear Stipulation Act. This law imposes stringent rules regarding the long-term management of highly radioactive wastes from nuclear power plants. It demands that a method for the management of these wastes be presented prior to the granting of permission to load and operate any new reactors in Sweden.

In December 1977 the Swedish nuclear industry presented a plan for the management of vitrified liquid wastes resulting from reprocessed spent fuel. The method was developed by the Nuclear Fuel Safety (*Kärnbränslesäkerhet*, or KBS) project. The KBS report was submitted to the government together with a request for permission to load and operate a new reactor, unit 3 at the Ringhals site. The decision on the Ringhals 3 application was the first under the 1977 Stipulation Act.

This report was prepared at the request of the minister of energy. It is a summary with special emphasis on safety analysis, including analysis of the importance of the different structures that should contain the wastes.

Nuclear power has become a political question. Thus radioactive waste management is subject also to political considerations. Such considerations are outside the scope of this work, which is limited to scientific and technical questions and their interaction with values.

The original Swedish edition of this report was published in August 1978 as *Kärnkraftens radioaktiva avfall: Inför Ringhals 3-beslutet*, DsI 1978:35 (Liberförlag) 1978. An English edition was also issued (DsI 1978:36). This book is based on that translation but has been adapted for an international readership by including a chapter on its Swedish background and by adding some material to the first three chapters. This edition, however, includes everything that was in the Swedish edition.

We would like to acknowledge the assistance of a number of other persons, among them B. Grundfeldt (Kemakta Konsult AB), fil dr R. Bergman, U. Bergström and S. Evans (Studsvik Energiteknik AB), G. Lind and M. Grill (Ministry of Industry), Professor E. Arrhenius (acting chairman of the Energy Commission Expert Group on Environment and Safety), Professor D. Abrahamson (consultant to the minister of energy), Dr. K. Edvardson (Institute of Radiation Protection), Dr. P. Hofseth (consultant to the Energy Commission Expert Group on Environment and Safety), Dr. B. Kjellström (AB Fjärrvärme), Professor B. Lindell (Institute of Radiation Protection), Dr. N.A. Mörner (University of Stockholm), Professor T. Westermark (Royal Institute of Technology), and Professor G. Östberg (University of Lund).

While gratefully acknowledging the assistance of these persons, we are responsible, of course, for any faults that may occur in this report.

T. B. Johansson and P. Steen

June 1980
Lund and Stockholm

PART I

Introduction and Summary

1

Background: Nuclear Power and Nuclear Waste Management in Sweden

During the early 1970s, the debate over the nuclear power program in Sweden focussed on nuclear waste management. The nuclear power issue then became a prominent part of political life and contributed to the fall of two governments. In early 1980 a national referendum advised that Sweden stop further orders for new nuclear power plants and phase out nuclear power entirely over the next twenty-five years.

In the 1950s Sweden had begun importing large amounts of oil. A government committee, already worried about oil dependence in the early fifties, suggested efforts for efficient use of energy. But by the mid-fifties atomic power's promise of abundant and cheap energy had dazzled Sweden. A major nuclear program, which helped defuse the nuclear weapons controversy in the late fifties and which took pressure off several major rivers threatened by hydropower development, was started.

The atomic energy program included education, research, and development. It was intended initially for hot-water production for district heating, but later it became oriented toward large light-water reactors producing electricity only. The vendor Asea-Atom, half-private and half-government owned, developed its own design.

The program started with Ågesta, a 10 megawatt (MW(e)) cogeneration station south of Stockholm, which operated from 1963 to 1974, when it was shut down for economic reasons. The first large electric station, the 460 MW(e) Oskarshamn I, was ordered in 1965 and became operable in 1972. Around 1970, Sweden had several units on order, and there was a general consensus that nuclear power was to grow rapidly, first, to satisfy the anticipated fast growth in electricity demand, then, to help replace imported oil. Sweden had the world's most ambitious nuclear program in the early 1970s. Before the first unit became operable, the Central Electricity Planning Board envisioned twenty-four units by 1990. Breeder reactors were considered as the next step and plans were made for the complete fuel cycle.

By 1973 eleven reactors, totaling 8,400 MW(e), had been licensed at four sites. Nuclear power was established.

The nuclear power decision had been taken without controversy. But dissent was to replace consent. It started in 1973. The opposition in environmental circles was growing, leading the way for the Center Party's shift toward an antinuclear position. Questions were asked in Parliament about reactor safety and waste management.

The resulting actions in Parliament led to a moratorium on nuclear power expansion pending further national debate. No new reactors would be licensed until more information became available on reactor safety and on waste handling. Parliament thus requested a national plan for nuclear power and also, in effect, assumed from government the ultimate responsibility for nuclear power.

By this time nuclear power had been withdrawn from its noncontroversial niche, had been placed squarely on the political agenda, and had attracted mounting interest in various quarters.

A political consequence of the oil embargo in 1973 and the subsequent rise in oil prices was that the decision on nuclear power had to be broadened into a decision on energy policy in general.

A debate of unparalleled intensity began during the fall of 1974. Not only was nuclear power among the topics but also energy supply alternatives, conservation possibilities, energy forecasts, the relationships between energy and real income, employment, and environmental issues. To some extent the debate became one about industrial society per se.

The government presented a bill on energy policy, which was passed by Parliament in May 1975. Its main thrust was to ensure the supply of energy until 1985 while preserving as many options as possible for the future. The major points were:

A target was to be established to reduce growth in annual energy demand from a historical rate of over 4 percent to no more than 2 percent until 1985 and a "serious inquiry" was to be made into the possibilities of achieving zero energy growth from 1990 onwards.

This was to be done through conservation measures, primarily within industry and space heating, stimulated by the price increase in imported oil and helped by government subsidies and cheap loans.

The nuclear program was increased from eleven reactors to thirteen to be completed by 1985 (five reactors were already operating).

A major new review of energy policy was scheduled to take place in 1978.

By this time nuclear power was a top item of political debate. Of four opposition parties, two nonsocialist parties backed the Social Democratic government in principle. One, the Center Party, the largest opposition party, would not accept any nuclear reactors in addition to the five already operating and called for more conservation. The fourth opposition party, the Communists, also rejected nuclear power and called for a referendum.

In April 1976 the final reports from a special investigation of spent nuclear fuel and other radioactive wastes were published.*

Several important decisions were postponed by the government, and the momentum of the nuclear program was thus slowed. No new sites for reactors were proposed, and the questions of waste handling and reprocessing were also postponed. So, too, was a discussion of the more and more sensitive question of mining Sweden's considerable uranium deposits.

By now energy policy had become a highly political issue, which it had never actually been since the introduction of hydropower at the beginning of the century. The Social Democratic government was voted out of office in 1976, partially because of its nuclear program. The new government, the first nonsocialist government in forty-four years, was split apart over nuclear energy, with one party opposed and the other two in favor. One of the opposing parties, the Center Party, gained support on a promise to phase out nuclear energy by 1985. The two others, the moderates and the liberals, were both supporters of the nuclear program. The new majority parties were divided hopelessly on energy policy in general and on nuclear policy in particular, but they managed to form a new coalition government. The way out had two major components, both postponing major decisions. One was to appoint the Energy Commission, which began its work in early 1977 and was expected to complete its report by spring 1978. This was to be the forum where the comprehensive energy program—discussed but not described during the 1976 campaign—would be developed. The other was the Stipulation Act, requiring that reactor licenses be granted by the government only after each licensee had come up with an acceptable scheme

*The reports of the AKA investigation are *Använt Kärnbränsle och radioaktivt avfall*, Part I, *SOU* 1976:30; *Använt Kärnbränsle och radioaktivt avfall*, Part II, *SOU* 1976:31; *Använt Kärnbränsle och radioaktivt avfall*, Appendices, *SOU* 1976:41.
The first volume has been published in English also as Spent Nuclear Fuel and Radioactive Waste, *SOU* 1976:32. *SOU* reports are published by the ministries and are available through Liber Distribution, Forlagsorder, S-162 89 Vällingby.

for waste management. In April 1977 Parliament passed the law. This law was seen by the pronuclear parties as a way to force the utilities to put forward a scheme for waste disposal; they had no serious doubts that the government would be able to accept the proposal rather soon and thus grant the licenses. The antinuclear party, however, saw the law and its safety conditions as an insurmountable barrier to the further expansion of nuclear energy. The precise conditions under which the proposals should be accepted or rejected were never spelled out in the law.

The Stipulation Act requires that prior to the operation of any additional nuclear reactor in Sweden, the reactor operator:

> Shall have presented a contract that adequately provides for the reprocessing of spent fuel and also shall have shown how and where the highly radioactive waste resulting from reprocessing can be deposited with absolute safety, or

> Shall have shown how and where the spent, but not reprocessed, nuclear fuel can be finally stored with absolute safety.

When the Stipulation Act was passed, Sweden had six reactors on line (Barsebäck 1 and 2, Oskarshamn 1 and 2, and Ringhals 1 and 2). Barsebäck 2 had been permitted to begin operation under a special section of the Stipulation Act. Four other reactors were under construction (Ringhals 3 and 4 and Forsmark 1 and 2). Another one of the thirteen reactors authorized by the 1975 energy decision (Forsmark 3) was not yet under construction, but some of the major components were being manufactured.

In December 1977 the Swedish State Power Board (Vattenfall) submitted an application under the Stipulation Act for permission to load fuel and to begin operation of the Ringhals 3 reactor. Vattenfall elected to make application under the Stipulation Act reprocessing option. Included in the application were (1) a contract between the French state company COGEMA and the Swedish Nuclear Fuel Supply Company (SKBF) for reprocessing the spent fuel from Ringhals 3 and (2) a report from the

Nuclear Fuel Safety Project, KBS-I. Vattenfall alleged that the reprocessing contract and the KBS-I report demonstrated that the demands of the Stipulation Act had been satisfied.

The KBS method, according to the usual practice, was reviewed by a number of Swedish organizations and governmental agencies. In January 1978 the Swedish government took the unusual step of requesting that a number of non-Swedish organizations and individuals participate in the review of the KBS method.

After intensive negotiations between the coalition parties, the government reached a decision on October 5, 1978, on the application for permission to operate the Ringhals 3 reactor under the terms of the Stipulation Act:

> The government has come to the conclusion that some further geological investigation will be necessary before the requirements of the Act can be considered entirely satisfied. . . . the law must be taken to imply that the applicant shall prove that an area, or areas, exist in Sweden which are of such nature that final storage in compliance with the requirements put down in the Act is possible. . . . Therefore, the complementary geological investigation ought to substantiate that there exists a sufficiently large rock formation at the required depth and with qualities that the KBS safety analysis, in other respects, gives as necessary prerequisites. . . . On the grounds of what has been said above, the application *cannot be granted* for the present (emphasis added).

It appeared, for a few days, that the government had survived its hardest test so far, but the continued nuclear construction program had not been dealt with.

Forsmark 3 was under construction. The estimated demand for electricity had significantly decreased since the reactor had been ordered, and no one, not even the utilities, seriously argued that the power generated by Forsmark 3 would be needed when the plant was completed. But the nuclear industry vigorously argued that there was a need for the jobs it would provide, and that the Forsmark 3 order was vital for the Swedish reactor-manufacturing industry. All parties agreed that stopping Forsmark 3 would be a decision to end domestic reactor manufacturing.

Complicating the situation was a violent attack in the press and by environmental groups against Prime Minister Fälldin and other Center Party ministers.

The government then resigned in October 1978, and the liberals (with only 39 out of 349 seats in Parliament) formed a minority government.

The Stipulation Act decision in October 1978 called for further investigations. These were performed during the fall of 1978; the results were submitted in February 1979.

The State Power Board report on these additional geological investigations claimed that the drilling at Sternö in southern Sweden had shown that a suitable site existed. The Swedish Nuclear Inspectorate was asked for advice by the government and appointed a panel of eight Scandinavian geologists to review the new KBS findings and to report to the inspectorate.

In March 1979 this panel of geologists submitted its report. Seven of the eight inspectorate-appointed reviewers stated: "The review group concludes that the Sternö region cannot be used for the repository proposed by KBS and cannot be used as a reference area for a final repository."

With circular reasoning, the inspectorate then argued that as long as the other containment structures were sound, the rock was not that important anyway. The inspectorate majority recommended that the government permit loading of Ringhals 3. This was granted in June 1979.

The KBS published a second report, called KBS-II in June 1978. It analyzes a final storage concept based on a direct deposition of the waste, without reprocessing. It has been the subject of a review but has not been the basis of any government decision under the Stipulation Act, or any other law.

After the June 1979 decision, and the Three Mile Island accident, nuclear waste issues have had lower visibility in Sweden.

The research program is rather extensive, however, and a site search is underway. In the spring of 1980, a decision was made to build an away-from-reactor storage facility for spent fuel.

The accident at Three Mile Island occurred two days after the inspectorate advice. This forced the Swedish pronuclear parties to agree to a nuclear referendum, long sought by the antinuclear parties, or to face the nuclear issue again in the September 1979 general election. They opted for the referendum, which was held in March 1980.

Three alternatives were on the ballot. Options one and two both stated that the six operating reactors and the four that were ready to operate should be allowed to operate for their economic life. The two that were under construction also were approved for operation for their economic lifetimes. This was estimated to be about twenty-five years. No additional reactors would be built.

One significant formal difference between options one and two was that option two included full nationalization of reactor operation.

Option three stated that the six operating reactors should be shut down within ten years and that the other six reactors should not be allowed to operate. Option three received 38.7 percent of the vote.

During the prereferendum campaign, the supporters of option one did not make a strong commitment to halt reactor construction after completion of the two that were being built. Option one was interpreted by many voters as being an open-ended nuclear option. It received 18.9 percent of the votes.

The supporters of option two, by contrast, strongly emphasized that the existing twelve reactors would be the end of the nuclear program. To emphasize its thrust, they presented a detailed thirty-seven-point program to demonstrate how this would be done. Option two received 39.1 percent of the vote.

The three options were actively supported by the five political parties that are represented in Parliament: option one by the moderates, option two by the liberals and the Social Democrats, and option three by the Center Party and the Communist Party.

The three nonsocialist parties in the coalition government each supported different referendum options.

A widely accepted interpretation of the referendum is that the 18.9 percent of the voters who supported option one wanted an open-ended nuclear program. The large majority of voters, the 77.8 percent who supported options two and three, advised that future energy supply would come, not from nuclear power, but from energy captured through hydropower, wind power, biomass, solar heating, and other solar technologies. Sweden already gets over 20 percent of its energy from hydropower and biomass, and several additional biomass systems are under development or construction.

Parliament has now acted on the outcome of the referendum. There are to be no further orders for nuclear power plants. Efficient use of energy and sustainable, preferably domestic, energy sources are to be the cornerstones of the future energy policy.

Sweden is thus the only country so far which has fully debated the solar versus nuclear issue and submitted the options to an informed electorate. Sweden has embarked on the route to replacing oil and nuclear energy over the next several decades with solar energy and more efficient energy use. It now seems that consent has replaced dissent. But energy policy and nuclear waste management will remain on the agenda for many years.

Further Readings

For further readings on the nuclear power and energy policy issue in Swedish debate the reader is referred to:

1. D. Abrahamson, "Governments Fall As Consensus Gives Way to Debate," *Bulletin of the Atomic Scientists*, November 1979; W. Barnaby, "First the Election and Then the Referendum," *Bulletin of the Atomic Scientists*, November 1979; W. Barnaby, "The Swedish Referendum: 'Do Away with It But Not Yet,' " *Bulletin of*

the Atomic Scientists, June 1980; D. Abrahamson and T. B. Johansson, "Summary of the Swedish Nuclear Referendum" (Report from the Hubert H. Humphrey Institute of Public Affairs, June 1980, reprinted by *Critical Mass Journal,* June 1980:10–11). Several reports and newsletters are available upon request from the Swedish Information Service, Swedish Consulate General, 825 Third Avenue, New York, New York 10022.

2. M. Lönnroth, T. B. Johansson, P. Steen, "Sweden beyond Oil: Nuclear Commitments and Solar Options," *Science,* May 9, 1980; M. Lönnroth, T. B. Johansson, P. Steen, *Energy in Transition,* Los Angeles: University of California Press, 1980; T. B. Johansson and P. Steen, *Solar Sweden: An Outline to a Renewable Energy System,* Stockholm: Secretariat for Future Studies, 1977; *Ambio* 7, no. 2 (1978); *Bulletin of the Atomic Scientists,* October 1979; M. Lönnroth, T. B. Johansson, P. Steen, *Solar Versus Nuclear: Choosing Energy Futures,* Oxford: Pergamon Press, 1980.

2

Introduction

The Swedish nuclear industry under the Stipulation Act has applied for permission to load and operate two reactors, Ringhals 3 and Forsmark 1. The application includes contracts for reprocessing with the French state company COGEMA and a report from the Nuclear Fuel Safety Project (KBS). In short, the KBS report alleges that a completely safe method for the final storage of high-level radioactive waste has been found, and therefore the requirements of the Stipulation Act have been satisfied.

Two distinctly different steps are required in the determination of whether or not the Stipulation Act has been satisfied. The first, which must precede the second, is to submit the application for technical review so that it can be determined whether the information contained therein is correct and complete. The second is the decision, which must be made through the political process, whether or not the method satisfies the applicable ethical criteria.

The KBS report has been reviewed through a domestic and international review process. This report attempts a concise summary of the information relevant to the safety of the proposal. It does not include a political analysis of the options open to the government in its consideration of the Ringhals 3 application. Nor does it include a recommendation as to whether or not that application should be approved or denied. The reason for this is that a recommendation is composed of two parts. One is the technical knowledge relevant to the proposed repository. The other is whether the resulting radiation exposures are found to be acceptable or not in the context of the people concerned and the time perspective involved. The latter is a value judgment and should, in our opinion, be separated from the technical question and handled through the political process using the best available information. The report does, however, include a summary and analysis of the relevant technical information currently available.

In order to make a judgment according to the Stipulation Act all relevant factors and consequences must be considered. It is not uncommon in theoretical studies to discover that factors or consequences that afterwards turn out to be important have been omitted or overlooked. It is, therefore, natural to focus on factors and consequences brought up in the review procedure which differ from the discussion in the KBS study.

For all problems under discussion there are two important questions: Are the factors and consequences treated by KBS regarded as correct, and have all relevant factors and consequences been identified?

The reviews of the KBS report include three general classes of material: descriptions of the KBS method (essentially repeating portions of the KBS reports); statements and opinions that are not supported by presentation of the evidence upon which that opinion is based; and presentation of evidence regarding the validity or completeness of factors included in the KBS reports, sometimes followed by the reviewers' conclusions and sometimes not.

To evaluate the accuracy and completeness of the KBS method, it is necessary to focus attention on the scientific evidence. It might appear to some readers that this report stresses critical comments or that it is not balanced. As with other scientific criticism, the reviewers of the KBS reports discussed in detail those points that they felt were subject to doubt and tended not to emphasize those things with which they either were in agreement or had not the qualifications to judge. Further, it is usually the critical questions that provoke consideration of weak steps in an argument.

There has been a large number of critical comments from the reviewers. What must be done is to form a judgment as to whether or not these comments are crucial and possible grounds to reject the KBS claims of absolute safety. To assist in this process, a sensitivity analysis has been done. The purpose of the sensitivity analysis is to demonstrate the exposures to radioactivity that result when various input parameters are varied within the range in which those parameters are uncertain. The same models used by KBS were used for the sensitivity analysis and the same analysts were employed. The Swedish way to review policy before government decision is to use the "remiss" procedure. This gives a combined policy review and technical review, and consequntly the comments offered are not always clearly separated.

Hearings and a formal review process such as those employed in some countries, for example the United States, are not used in Sweden. In the Swedish remiss process the major documents upon which a governmental decision is to be based are sent to private and public organizations and agencies for comment. These will include the governmental agencies that have specific regulatory functions—for example in the Ringhals 3 case the Nuclear Inspectorate (SKI) and the Institute of Radiation Protection (SSI).* Universities, labor organizations, professional organizations, the learned academies, and many other groups may also be asked to comment. Each of these organizations may use whatever internal processes they wish in carrying out their reviews. The result of the remiss is a set of written comments

*For list of abbreviations see chapter 17.

from the participating organizations. These comments are sent to the government ministry that has the leading role in the decision in question. (In the case of the remiss on the Ringhals 3 applications, it was the Ministry of Industry.) The ministry staff then summarizes the comments and forwards the summary to the government. There is no formal process to assure that the review is complete, nor is there a formal process to establish the reasons should different review organizations include apparently conflicting information in their reviews.

The Ringhals 3 application was reviewed by twenty-four Swedish organizations during a remiss process.* Each organization commented on the KBS-I proposal. The COGEMA/SKBF reprocessing contract is secret and so was available only to very few agencies.

In addition to the Ringhals 3 remiss, the KBS-I report was subjected to technical review. The Safety and Environment Expert Group of the Swedish Energy Commission conducted a limited technical review.** Because such a large number of the qualified Swedish scientists and engineers had participated in the KBS work, and because of the extraordinary seriousness with which the management of radioactive wastes is viewed, the Swedish government decided also to sponsor an international review to supplement the usual Swedish remiss. Twenty-four organizations or individuals submitted comments during the foreign review,*** either general reviews of the entire KBS scheme or special reviews of specific portions of the KBS-I report.

Neither the KBS-I report nor other Swedish studies include a thorough analysis of the economic costs associated with the

---

*The Swedish remiss answers are published as "*Yttranden över statens vatten-fallsverks ansökan enligt villkorslagen om tillstånd att tillföra reaktoranläggningen Ringhals 3 kärnbränsle*", *DsI* 1978:29.

**The Energy Commission review is published as "Disposal of High Active Nuclear Fuel Waste", *DsI* 1978:17 (in English).

***The international review is published as "Report on Review through Foreign Expertise of the Report 'Handling of Spent Nuclear Fuel and final Storage of Vitrified High level Reprocessing waste,' " *DsI* 1978:28 (in English).

management of radioactive wastes, nor do they include a description of a management program to coordinate and assume responsibility for the research, development, and demonstration that must be done prior to the implementation of any waste storage program. Therefore, this report makes virtually no comment on economic or on programmatic matters.

3

Summary

The Swedish State Power Board (*Statens vattenfallsverk*) has made application to the government for permission to load a nuclear power reactor under the 1977 Stipulation Act.* The application includes:

1. A report from the Nuclear Fuel Safety Project (*Kärnbränsle-säkerhet*, or KBS),** which describes a method for handling the vitrified liquid high-level waste from reprocessing.

2. Contracts between the Swedish nuclear fuel company

*Stipulation Act (*Villkorslagen*) 1977:140, *Prop* 53(1976–1977).

**There are two reports from the KBS project. KBS-I, published in December 1977, describes a method for handling of vitrified liquid reprocessing wastes. It is the KBS-I report that is contained in the Ringhals 3 application and that is dealt with in this review. The second KBS report, KBS-II, was published in June 1978 and describes a method for handling unreprocessed spent fuel.

(*Svensk Kärnbränsleförsörjning AB*, or SKBF) and the French state company COGEMA for reprocessing of Swedish spent fuel at the UP3-A plant at La Hague.

The Ringhals 3 application has been approved according to the 1956 atomic energy law* and the 1958 radiation protection law.** The pending decision on operation of Ringhals 3 is whether or not the demands of the Stipulation Act have been satisfied.

The Ringhals 3 application has been subjected to scientific review. The Swedish remiss process included review of both the KBS-I report and the reprocessing contracts. The international review was restricted to the KBS-I report. The results of these reviews must be examined in the context of the demands of the Stipulation Act.

The Stipulation Act, in the clause referred to by KBS, requires:

That there be an *acceptable* reprocessing contract

That it be shown *how* the highly radioactive waste resulting from reprocessing can be finally stored

That the waste be handled in a way that is *absolutely safe*

That it be shown *where* the highly radioactive waste resulting from reprocessing will be placed for final storage.

The Stipulation Act allows for two options: with or without reprocessing. The decision to allow reprocessing of the fuel involves not only economic and waste management considerations but also political factors associated with potential interrelationships between civilian and military applications of nuclear energy and other plutonium problems. The Ringhals 3 application was made under the reprocessing option. A final storage of the waste assumes that reprocessing will take place. If this cannot be done the final storage of the waste cannot be carried out according to the KBS-I method.

*Atomic Energy Act (*Atomenergilagen*) 1956:306.
**Radiation Protection Act (*Stralskyddslagen*) 1958:110.

Both Scientific and Political Analyses Are Needed

Some of the issues involved in making the Ringhals 3 decision can reasonably be answered through technical scientific inquiry. Others can only be solved through the political process.

This book attempts to summarize the extent of scientific knowledge, as represented by the technical reviews, with special emphasis on questions that are important for the safety analysis of the method that Vattenfall proposed in the Ringhals 3 application.

There is no precedent upon which to draw for operational meanings of "has been shown," "highly radioactive waste resulting from reprocessing," "where," or "absolutely safe" in the context of the Stipulation Act. During the reviews several possible meanings of these, and other, terms were suggested. What demands will be placed on the quality of technical and scientific evidence is a central question that must be settled by the responsible authorities and political officials.

Results of the Review

Reprocessing Contracts

The application includes two reprocessing contracts between SKBF and COGEMA. The remiss responses of SSI and SKI, based on the April 1977 contract, found the contract unacceptable as it did not provide for reprocessing of all fuel that would be made radioactive during initial operation of Ringhals 3.

The March 1978 contracts provided for reprocessing of fuel from several reactors, including Ringhals 3, during the 1980s, and SSI and SKI found that the contracts now satisfied the stipulated demands.

At present the contracts are secret. For that reason this

summary mentions only those portions of the contracts that earlier had been made public.

In a letter to Ministry of Industry SSI's chairman, professor Bo Lindell, stated that the contract did not fulfill the demands of the Stipulation Act as it did not provide a guarantee that reprocessing would actually take place.

In his reservation to the SKI remiss answer Thomas B. Johansson noted that the force majeure clause permitted COGEMA to forego reprocessing for a number of reasons, including technical difficulties of a radiological nature. He also pointed out the very limited world experience in reprocessing high-burnup LWR fuels and the radiological difficulties that have been encountered in those reprocessing plants that have operated in the past. It is therefore uncertain as to whether the planned facility could be operated as anticipated.

Several reviewers, including FOA and SSI, expressed concern that the carbon-14 produced during reprocessing would continue to be released into the environment. Were that done, the SSI radiation exposure guidelines might be exceeded. No reviewer concluded that this omission was cause to find that the contract did not satisfy the demand of the Stipulation Act.

The KBS-I Method: General Comments

The general scheme for the KBS-I method for handling and final storage of high-level waste is: removal of spent fuel from the reactors and temporary storage at the reactor site; transportation of the spent fuel to a central spent-fuel storage pool (*centrallager*) where it would be further stored; transportation to France; reprocessing by COGEMA at the yet to be constructed UP3-A plant at La Hague; vitrification of the liquid high-level waste; return of the vitrified waste to Sweden; storage of the vitrified waste in a transitional storage facility (*mellanlager*); encapsulation of the glass in a lead and titanium capsule; placement of the waste into a final repository located 500 meters deep in granite.

No reviewer argued that major steps necessary for the handling of vitrified, highly radioactive waste, as described in the KBS-I report, had been omitted.

The KBS-I method does not include a description of the handling of the plutonium that would be separated as a result of reprocessing.

"Absolute Safety" of the KBS-I Method

The Stipulation Act requires that highly radioactive wastes be managed in a way that is absolutely safe.

The KBS-I report includes a safety analysis of the proposed method for handling the vitrified, highly active waste resulting from reprocessing. That method includes several steps between the reactor and the final repository.

The KBS-I report describes a centrallager, transportation, reprocessing and glassification, and a mellanlager. The assumption made by KBS is that given the present state of experience with other nuclear facilities, it is reasonable to assume that these facilities can be constructed and operated in a manner consistent with appropriate regulations and standards. Although many reviewers pointed out potential problems, no reviewer assumed that the demands of the Stipulation Act had not been fulfilled.

The detailed safety analysis in KBS-I was confined to the long-term performance of the proposed final repository. Using their estimates of conservative values for the parameters describing the proposed repository, and their mathematical models, KBS concluded that it is unlikely that the resulting radiation exposures to the critical groups of persons living near the repository would exceed 13 millirems per year. The maximum radiation exposure would not be reached for approximately 200,000 years.

No activity can be absolutely safe. There must be an operational definition. It is usual practice to define *safe* in nuclear activities by reference to the appropriate radiation standards.

No standards for permissible radiation exposures from the

final storage of high-level radioactive waste have been established. However, Swedish authorities have published specific standards for the operation of reactors and also general standards for the complete nuclear fuel cycle including the waste portion.

The highest future dose to which any individual is exposed in a critical group must not exceed 50 millirems per year and preferably should not exceed 10 millirems per year if present stipulations for radiation doses around nuclear power plants are to be fulfilled. The International Commission on Radiological Protection (ICRP) dose limits for individuals receiving high radiation doses during a succession of years is 100 millirems per year.

Theoretical calculations can also be made of the total future radiation doses from a repository of radioactive waste to a world population over long periods (hundreds of thousands or millions of years). These radiation doses can be converted into a number of future deaths caused by final deposition of radioactive waste. This is discussed in chapter 14.

KBS-I claims that the values that have been chosen in its safety analysis are conservative. As discussed in detail in chapter 13, this claim is challenged by many technical reviewers. Parameters for which the KBS-I values may be not at all conservative include the glass-leach rate and the transport time for groundwater from the repository to the surface.

The reviewers do not, of course, all agree on what is the most likely value or what values should be chosen in order to cover the uncertainties in the records. Such is never the case in situations where there is scientific uncertainty.

KBS failure to consider the wastes produced by a reactor that uses plutonium as nuclear fuel also introduces a nonconservative element into the KBS safety analysis. The Ringhals 3 application was made under the reprocessing option of the Stipulation Act. The only purpose of reprocessing is to recover the plutonium from the spent fuel. The only civilian use for the recovered plutonium is as fuel in reactors: either in light-water reactors, such as Ringhals 3, or in breeder reactors. A reactor fueled with plutonium produces radioactive wastes containing considerably more radioactive, long-lasting, transuranic materials than does a reactor fueled with virgin uranium. It is these transuranic nuclides that would be responsible for most of the human exposure to radiation according to the KBS-I method. Yet, the KBS safety analysis does not use a nuclear fuel containing plutonium and uranium. This omission has been pointed out during the review, and calculations of nuclear fuel containing plutonium and uranium have been carried out in this report.

The Result of the Sensitivity Analysis

The purpose of a so-called sensitivity analysis (described in chapter 14) is to give an understanding of how variations in the numerical parameters that describe how radioactive material can escape from a repository would change the radiation doses to man emanating from the repository. This analysis used the same models that were used by the KBS-I project. The actual computations were done by the same persons who did the computations for the KBS-I project.

The size of the releases of radioactive materials that can reach the surface and the biosphere from a repository depends on several factors. Those of importance are shown in table 3.1 along with figures partly stated by KBS to be conservative values (KBS main case) and partly with figures from two sensitivity analyses. The resulting radiation doses are stated in the table and in figure 3.1.

Each of the input values in the safety analysis is within the range of scientific uncertainty. It is not claimed that they are the most probable values, but considering the review, they cannot be excluded with the present state of knowledge. The resulting radiation doses amount, in the case where uranium is the fuel, to about 100 millirems per year both in the well alternative and in the lake alternative with the difference between those two alternatives being small.

The uncertainties of the consequences of a final deposition can be seen in table 3.1. To those uncertainties should be added

TABLE 3.1

RADIATION DOSES TO INDIVIDUALS FROM THE
KBS-PROPOSED REPOSITORY

	KBS main case		Sensitivity analysis	
	Uranium fuel[1]	Reprocessed plutonium fuel[2]	Uranium fuel[3]	Reprocessed plutonium fuel[4]
Parameters				
Time for capsule disruption (years)	1,000	1,000	500	500
Glass-leaching time (years)	30,000	30,000	6,000	6,000
Groundwater transport time (years)	400	400	40	40
Retention of radioactive materials in the bedrock		same in all cases		
Resulting radiation doses Maximum individual dose to critical group in a future situation when a well is drilled in the vicinity of the repository, millirems/years[5]	13 mrem/year	33 mrem/year	100 mrem/year	1,000 mrem/year

1. KBS main case (virgin uranium as fuel)
2. KBS main case (using recycled plutonium fuel only, no other changes). This case was not discussed by KBS but was calculated as case D 1 in the sensitivity analysis.
3. Case B 1 in the sensitivity analysis.
4. Case B 2 in the sensitivity analysis.
5. Millirem = mrem = one-thousandth rem. Rem is a measure of radioactive dose.

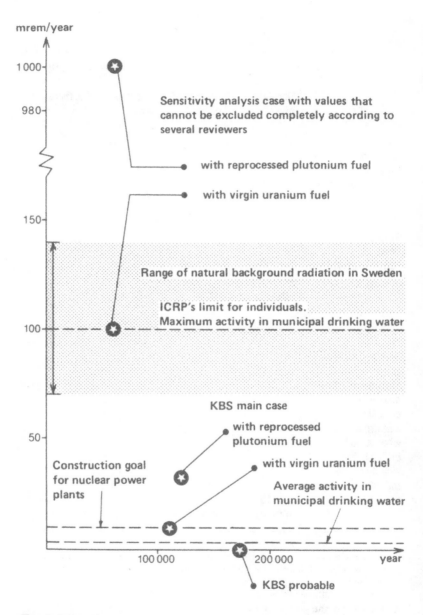

Fig. 3.1 Maximum annual radiation doses to individuals (well alternative).

the uncertainties of the models used when calculating the radiation doses. Models are always simplifications of reality, and their reliability depends among other things on the input data.

If some of the data used in the models were changed to reflect existing uncertainties, the radiation doses might, in disadvantageous cases, be more than ten times higher than stated in table 3.1 and figure 3.1. In KBS main case, for example, this would mean radiation doses exceeding 100 millirems per year in the well alternative.

The most important areas where KBS assumptions about the final deposition were questioned by the reviewers are briefly described below. In certain cases several reviewers considered the scientific basis inadequate to support KBS's claim that the values that were chosen in the safety analysis cover the range of uncertainty.

WASTE GLASS

The leaching of glass and radioactive materials from the repository depend among other things on the glass properties, the amount and composition of the groundwater, the temperature, the amount and composition of the wastes, and the pressure of ionizing radiation. Systematic studies of all variables and their united actions are very limited and have only been made in laboratories.

> The leach rate of the vitrified waste into the groundwater used in the KBS analysis has not been proved to be a conservative assumption, according to the review.

> No experiments or measurements have been made with cylinders of the size proposed by KBS and containing the suggested quantity of radioactive materials.

> Stress in the glass cylinders might result in shattering. The area that might come into contact with the groundwater would therefore be much larger than supposed.

> In conclusion, it cannot be precluded that the time for a glass cylinder, which has been exposed to groundwater, to dissolve might be a fifth or less of the value used by KBS.

ENCAPSULATION

Titanium is a material that has not been used in large quantities until recent decades, and there are uncertainties about its behavior over thousands of years.

Mechanisms, such as hydrogen-induced embrittlement, might induce failure of encapsulation. Present knowledge is not sufficient, according to several reviewers, to dismiss completely the notion that failure might occur earlier than assumed by KBS.

ROCK PROPERTIES

The function of the various barriers, such as glass, encapsulation, and rock, is to delay the entry to the biosphere of the radioactive materials for as long as is necessary for them to decay into harmless elements.*

The third barrier, bedrock, is not a question of technique, that is, something that man shapes according to choice. The discussion of bedrock has three parts:

What properties must the rock have to be regarded as an acceptable barrier?

Does such rock exist?

Where can an actual site with these properties be found for final deposition which furthermore is acceptable from other points of view (e.g., not containing ore or other mineral deposits for future mining)?

A repository** requires an area of at least one square kilometer. Three areas have been investigated, and one to three boreholes have been drilled in each.

The KBS assumption of a groundwater transport time of four hundred years is based on measurements of the groundwater age with the carbon-14 method and theoretical calculations including measurements in boreholes.

---

*The radioactive materials in the waste decompose when emitting radioactive radiation (ionizing radiation). After a time (longer or shorter time depending on the material in question), the material has decayed to a stable element when radiation has ceased also.
**Containing wastes from thirteen reactors during their lifetimes.

The measurements that have been made of the ground-water age with the carbon-14 method are perhaps faulty. These faults are of such a magnitude that they cannot support the assumed groundwater transport times from a repository to a well or a lake. Measurements of groundwater age reflect the time for the groundwater to move from the surface to the repository.

The measurements that have been carried out have shown crush zones with a high water permeability even at great depths.

The assumptions of groundwater transport time are regarded as uncertain by several reviewers whose theoretical calculations, using measurement parameters, result in groundwater transport times between ten and a few hundred years.

No measurements have been made at depths below a repository where water-permeable and water-transporting layers might be located.

The measurements of average permeability do not represent the water flows through cracks.

It is uncertain how the existence of a suitable location with the desired bedrock could be demonstrated without destroying the area (for instance, through a large number of boreholes).

The retention of the radioactive materials as a consequence of various mechanisms relative to groundwater movement has only been measured in laboratories, and it is uncertain how well they show the conditions around a repository at five hundred meters depth in bedrock.

Filling and sealing of tunnels and shafts with bentonite clay is a method where the necessary impermeability has not been verified through practical experiments.

Where the Wastes Would Be Finally Placed

The Stipulation Act requires that it be shown where the highly radioactive wastes would be finally stored.

KBS-I does not claim that the place where the final repository will be placed has been located. KBS-I does claim that sufficient investigations have been carried out to show that it is reasonable to believe such a place does exist in Sweden, for example, in one of the three areas where test drillings have been carried out.

Nowhere in the world have siting criteria for a final repository been established. The parameters used for the KBS-I final repository safety analysis constitute an implicit set of criteria. It is reasonable to assume that a place could be considered acceptable *if* its characteristics are, in every significant respect, equal to or better than those assumed by KBS in the safety analysis, *if* the mathematical models used for the safety analysis adequately represent the actual physical situation, and *if* the resulting radiation doses calculated from those models, using the parameters that describe the actual situation, are acceptable.

The issue is whether or not an actual block of rock exists in Sweden which fulfills the demands and which could be located. Many reviewers assume that it should be possible to locate a suitable place for a final repository. Others are not convinced. All, however, seem to agree that it has not yet been shown that an appropriate place actually does exist.

Concluding Comments

The two major judgments that have to be made are how to interpret the concepts "has shown" and "absolutely safe" as used in the law. The law does not contain explicit criteria to guide this interpretation. In an attempt to connect values and scientific-technical information we illustrate some combinations of conceivable combinations of these concepts. In figure 3.2, we have formulated three different steps on the continual scale of possible interpretations. Thus we obtain a three-by-three matrix that can be filled in with *yes* or *no* according to the status of the technical and scientific problem to be dealt with. Cells of the matrix are filled in based on the evidence presented. What should be re-

INTERPRETATION OF "ABSOLUTE SAFETY"

Answer to the question: "Has it been shown that the KBS-I final storage method is absolutely safe?"		Nobody should be exposed to radiation doses that are not accepted today within the next 1,000 years.	Nobody should be exposed to radiation that is not accepted today within 1 million years.	Future radiation doses from a repository should be substantially less than the doses accepted today, as a hedge against today's lack of knowledge.
INTERPRETATION OF "HAS SHOWN"	Some experts believe that the specified conditions can be fulfilled in the future.	Yes	Yes	Yes?
	Experiments have shown that the specified conditions are fulfilled now.	Yes?	No	No
	It has been demonstrated that the conditions are fulfilled (e.g., by producing a full-scale canister and by building tunnel and shafts)	No	No	No

↑ Decreasing demand on "has shown"

← Decreasing demand on "absolute safety"

Fig. 3.2 Policy matrix: Does the proposed final storage method fulfill the Stipulation Act? Besides scientific-technical data, a decision is based on interpretations of the concepts "has shown" and "absolutely safe."

quired to fulfill the Stipulation Act is a matter of ethics and is thus a political decision. The matrix is intended to serve as a guide in political decision making, a way of trying to put the complex scientific-technical issue within reach of the political process.*

The development process is such that the extent of knowledge is constantly expanding. The KBS-I method is restricted to what is known today. There is no question but that future scientific and engineering developments will increase the available knowledge about the final storage of highly radioactive wastes.

Yet, the fact is that a decision must be made now on the Ringhals 3 application, and it must be made in the face of many uncertainties. Either a positive or negative decision could bring forth political conflicts. A negative decision carries with it the implication of economic penalties. But a negative decision today can be changed to a positive decision in the near future if ongoing research and development work brings forward hoped-for results. A positive decision now might be regarded by some as insupportable in the light of the stringent requirements of the Stipulation Act compared with the technical-scientific uncertainties of the consequences of a final repository.

*The matrix is a way of presenting the findings in our report. It was published in an article of ours in the Swedish newspaper *Dagens Nyheter*, September 27, 1978.

PART II

The Technical and Legal Settings for the Ringhals 3 Decision

4

The Legal Background

The Nuclear Stipulation Act

The Nuclear Stipulation Act[1] requires that, prior to the loading of fuel or operation of any additional nuclear reactor in Sweden, the reactor operator either:

1. will have presented a contract that adequately provides for the reprocessing of spent fuel and also will have shown how and where the final deposition of the highly radioactive waste resulting from the reprocessing can be effected with absolute safety, or
2. will have shown how and where the spent, but not reprocessed, nuclear fuel can be finally stored with absolute safety.

Application of the law to a specific case requires that the demands of the Stipulation Act be defined.

The Stipulation Act allows for two options in the back end of the nuclear fuel cycle: with or without reprocessing. There is no indication that the intent was that a Swedish decision on reprocessing should be made in connection with an application for reactor operation under the Stipulation Act. Would the granting of permission to load and operate a new reactor based on an application assuming spent-fuel reprocessing also be de facto approval of reprocessing?

The detailed description of the Stipulation Bill, which was presented to the Parliament in the proposition,[2] stated that:

> The term *contract* implies that there be a legally binding agreement with such contents that it satisfies the existing demand for reprocessing of the spent fuel in the nuclear reactor. The agreement shall furthermore be made with someone who has the means for reprocessing and who otherwise can be expected to fulfill those demands that result from the agreement. . . . Should the period of validity of the agreement be shorter than the operating time of the reactor, that a time limit on the permission to run the reactor be posed must be considered so that it corresponds to the period of validity of the contract.

Apart from a contract, need it have been shown that reprocessing actually can be accomplished on a commercial scale? The French government has indicated that it will not consider any reprocessing contract valid until there has been a government-to-government agreement specifying certain matters not stated in the commercial contract. Does the Stipulation Act require satisfactory negotiation of such government-to-government agreements prior to having an acceptable contract?

What degree of experimental and practical verification does the term "has shown" demand? The proposition states that "it is not demanded that a site for deposition has been completed when an application is considered."[3]

In the past, and for purposes unrelated to those of the Stipulation Act, highly radioactive wastes have been defined as including only one of the several forms of waste from reprocessing. Several different definitions could be considered (see chapter 7).

There is a wide range of opinion as to the interpretation of "absolute safety." Is it sufficient to show that the waste management methods will, with reasonable assurance, comply with applicable standards and guidelines regulating exposure to ionizing radiation? (See chapter 8.)

The proposition to Parliament includes a statement that there can be no damage to the ecological system from any phase of handling the spent reactor fuel. Is this the standard that must be met? What is "damage"?

Does the demand "where" mean that it is sufficient that there are reasons to believe that at least one site, acceptable from a technical standpoint, does in fact exist somewhere in Sweden, or is it required that such a site actually be found and its suitability proved? Does it mean that political (for instance, application according to the Building Act § 136a) as well as technical obstacles must be overcome? (The question "where" is discussed in chapter 15.)

The question about the handling and final storage of the waste is discussed in chapters 9-13.

The Application for Operation of the Ringhals 3 Reactor

Several reactors were under construction in Sweden when the Stipulation Act was passed into law. Two of these, unit 3 at the Ringhals site and unit 1 at the Forsmark site, are now deemed to be complete and ready for operation.

In December 1977 the Swedish State Power Board (Vattenfall) submitted an application under the Stipulation Act for permission to load fuel and to begin operation of the Ringhals 3 reactor. Vattenfall elected to make application under the Stipulation Act reprocessing option. Included in the application were a contract between the French state company COGEMA and the

Swedish Nuclear Fuel Supply Company (SKBF) for reprocessing the spent fuel from Ringhals 3, and the KBS-I report. Vattenfall alleged that the reprocessing contract and the KBS-I report demonstrated that the demands of the Stipulation Act had been satisfied.

The Ringhals 3 contract for 20 tons of fuel was signed in April 1977. The reprocessing was to take place in the existing reprocessing plant at La Hague. A second contract between SKBF and COGEMA, for 620 tons of uranium, was signed on March 16, 1978. The reprocessing was to take place in the planned UP3-A plant at La Hague. This contract was added to the Vattenfall application for Ringhals 3 on May 18, 1978.

Negotiations between the governments of Sweden and France were undertaken in 1977. The necessary government-to-government agreement has not yet been concluded.

The Nuclear Fuel Safety (KBS) Project

In response to the Stipulation Act, those Swedish electric utilities that own or operate reactors jointly formed and wholly funded the Nuclear Fuel Safety (KBS) Project. The KBS project was charged with doing studies and preparing reports to support an allegation that the demands of the Stipulation Act had been satisfied.

The KBS project released its first report in December 1977. This report (KBS-I)* described a management proposal for the

---

*Handling of Spent Nuclear Fuel and Final Storage of Vitrified High-Level Reprocessing Waste, KBS project.

Volumes: *I. General*
            II. Geology
            III. Facilities
            IV. Safety Analysis
             V. Foreign Activities

Available from KBS, Brahegatan 47, S-102 40 STOCKHOLM
A large number of technical reports (KBS-TR) has also been published.

vitrified liquid, high-level radioactive waste that would result were spent reactor fuel to be reprocessed. The KBS method for management of the highly radioactive waste from the back end of the fuel cycle consists of the following steps:

1. Removal of the spent fuel from the reactors
2. An optional ten-year storage of the spent fuel in a facility yet to be built in Sweden, the centrallager
3. Transportation of the spent fuel to France for reprocessing
4. Reprocessing, followed by vitrification of the aqueous solvent extraction waste and encapsulation of the glass into stainless steel canisters
5. Beginning in 1990, at the earliest, the vitrified waste canisters shipped back to Sweden
6. Interim storage of the waste canisters for at least thirty years in a Swedish facility, the mellanlager having a 30-meter rock cover
7. Encapsulation of the canisters in a second canister consisting of an inner 10-centimeter lead layer and an outer 6-millimeter titanium can
8. Emplacement of the encapsulated waste in a final repository at a depth of approximately 500 meters in suitable crystalline bedrock formations in Sweden beginning no earlier than 2020.
9. Backfilling the storage holes and repository tunnels with a quartz sand and bentonite clay filler material.

KBS states that this design of the back end of the nuclear fuel cycle fulfills all the requirements set forth in the Stipulation Act for demonstrating how and where high-level radioactive wastes resulting from reprocessing can be stored with absolute safety. This conclusion is based on the following analysis:

1. Field investigations were carried out at five sites. These investigations indicated that suitable bedrock for a repository could be found in Sweden.
2. Three of the five sites were considered to offer satisfactory conditions for final storage. It was therefore concluded that

rock formations with equivalent conditions also were available at several other places in Sweden.

3. An extensive safety analysis concluded that the release of radioactive substances that could occur in connection with normal operation or with an accident in some of the stages of spent-fuel and waste handling within Sweden would be insignificant in comparison with corresponding conditions at a nuclear reactor.

4. Complete safety of final waste storage was considered to be assured by providing conditions that either prevent or delay the movement of radionuclides for a sufficient period of time during which the nuclides are rendered harmless before reaching the biosphere. These conditions were considered met by imposing multiple barriers to nuclide release. For example:

   a. The waste is protected by a stainless steel can and a titanium and lead outer capsule. The encapsulation is very resistive to corrosion.

   b. The waste form itself, glass, has a low leach rate even if the capsule were breached.

   c. The quartz sand and bentonite clay buffer material has a low permeability and high retention capacity for waste radionuclides, thus restricting water flow and delaying nuclide migration.

   d. The slow rate of movement of water that does manage to penetrate these barriers, coupled with the long distance to the biosphere, results in a very low rate of release of radioactive materials to the biosphere.

   e. Dilution of any released waste into large volumes of groundwater will take place before entry into the biosphere.

5. The worst-case analyses indicate that the maximum individual dose from repository failure could be 13 millirems per year and would not occur for at least 200,000 years. This dose is considerably lower than the dose, 100 millirems per year, recognized by the International Commission on Radi-

ation protection as the upper limit for permissible additional doses for individuals. The Swedish National Institute of Radiation Protection requires that expected additional doses for the critical group living in the vicinity of a nuclear power plant be no more than 10 millirems per year.

6. The assumptions and data used in the safety analysis were indicated to have been selected with safety margins. The actual dosage considered probable is about one percent of the maximum value of 13 millirems per year and therefore would be virtually insignificant.

5

The Nuclear Fuel Cycle

Radioactive Materials in the Nuclear Fuel Cycle

The nuclear fuel cycle includes all steps from the mining of uranium to the ultimate management of radioactive waste products (figure 5.1).

Radioactive materials are present throughout the fuel cycle. Although handling of radioactive material entails difficulties during all the steps of the fuel cycle, the radioactive materials (including plutonium) created during operation of the nuclear reactor are of importance in the context of the Stipulation Act.

During reactor operation fission products, new transuranic elements, and other radioactive isotopes are created. A small fraction of these radioactive substances are routinely released into the environment at power reactors during normal operation.

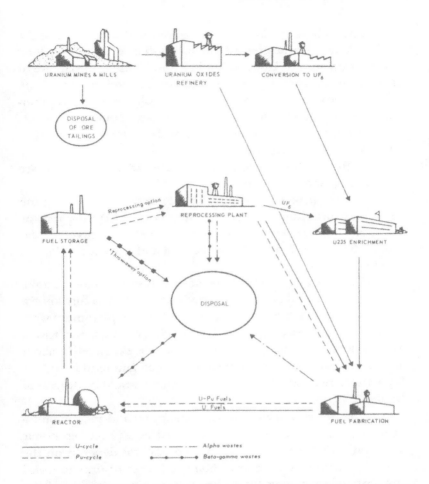

Fig. 5.1 Major steps in the nuclear fuel cycle and main radioactive waste (source: "Objectives, concepts and strategies for the management of radioactive waste arising from nuclear power programmes," OECD, NEA, September 1977, p. 26).

Most of the radioactive products are, however, retained in the used reactor fuel. A portion of this spent fuel is removed from each reactor annually and replaced with new fuel. The steps beginning with the removal of the spent fuel from the reactor and extending through all phases of spent fuel management are collectively called the back end of the fuel cycle.

The radioactive materials produced in the back end of the fuel cycle are characterized as being small in volume, very long lasting, and accumulating in food chains. These materials are of three general types: fission products, such as cesium, strontium, or iodine; transuranic materials, such as plutonium or neptunium; and activation products, such as the structural components of a reactor or a reprocessing plant that become radioactive because of exposure to radiation.

Within each of these categories are radioisotopes that are long-lasting and must be contained.

The purpose of reprocessing is to isolate and recover the remaining uranium and the plutonium created during reactor operation and remaining in the spent fuel. If recovered, the fissionable plutonium could be recycled and used as an alternative to uranium.

If only uranium were to be used as fuel in fission reactors, such reactors could be used for only a few decades before known sources of uranium would be exhausted. If the plutonium were recovered and used in so-called breeder reactors, the fission energy potential of natural uranium would be expanded by about sixty fold. If the plutonium were recovered and used in today's light-water reactors, uranium requirements would be decreased by about 30 percent.

Whether or not to reprocess spent fuel and recycle plutonium has, until recently, been regarded as a purely economic question. Whether reprocessing is economical depends on the cost of reprocessing, the availability and price of uranium, and the costs of managing the radioactive products resulting from reprocessing. These costs are not yet well known. In addition to these usual costs there are costs that are even more difficult to quantify but are nonetheless important. As stated in the Swedish paper to the 1977 IAEA Nuclear Power conference:

> There is also special concern relating to spent fuel, plutonium, and waste that is warranted by the particular characteristics of the materials, such as the extremely long-term hazards of plutonium and the significant risks for proliferation of nuclear explosive capacity connected with the re-

processing of spent fuel and the handling of plutonium. In this field the concern for safety, in the widest sense of the term, has placed demands on governments and the nuclear industry far beyond limits that in other contexts would be deemed reasonable.[1]

The Effects When People Are Exposed to Radiation

The individual can be injured by ionizing radiation in more than one way. High doses can cause death relatively soon after exposure. Doses of this magnitude can only be considered when related to serious accidents. The doses related to the normal handling of wastes are much smaller.

It is generally accepted, for purposes of considering standards governing exposure to ionizing radiation, that the late effects associated with exposures are: cumulative (doses received at different times being strictly additive); proportional to dose; and no-threshold phenomena (no dose below which there is no effect).

Ionizing radiation causes cell damage, both hereditary (affecting future generations) and carcinogenic (inability to control the growth in the cells of exposed individuals).

According to the Safety and Environment Expert Group of the Swedish Energy Commission, the value of the risk factor for genetic damage is thirty cases per 10^6 manrems for the first generation and 100 cases totally for all future generations. The cancer risk has been estimated to be between one hundred and one thousand cases per 10^6 manrems.

Whether or Not To Reprocess Spent Fuel

There are two basic options for the management of the spent reactor fuel: (1) the "once-through" alternative and (2) the "reprocessing and recycle" alternative.

These costs, both the usual ones that appear directly on the ledgers of the electric utilities and the extraordinary ones that arise because of the "particular characteristics of the materials," must somehow be balanced against the potential value of the plutonium.

There are therefore several interrelated questions that must be answered before knowing whether or not the ultimate waste from nuclear reactors is in the form of unreprocessed spent fuel (the once-through alternative) or the radioactive products resulting from reprocessing (the reprocess and recycle alternative). These include:

Is it technically possible to reprocess the spent fuel?

If so, is it economic to reprocess, particularly for a country having large domestic uranium resources?

If so, is it politically acceptable to reprocess, taking into account the public health and the weapons proliferation implications of a "plutonium economy"?

If so, would a decision to reprocess make it unnecessary to plan for the direct storage of unreprocessed spent fuel?

Radioactive Products Resulting from Reprocessing

Were the spent fuel reprocessed, the following radioactive products would result:

Plutonium and recovered uranium

High-level liquid wastes

Cladding hulls

Sludges and other solid or semisolid materials

Transuranic contaminated wastes from the reprocessing plant and also from subsequent processses, for example, fuel fabrication involving plutonium

Radioactive gases, primarily krypton-85, iodine-129, hydrogen-3 (tritium), and carbon-14 (as carbon dioxide)

Long-lived wastes from decommissioning of the reprocessing plant itself.

The wastes are all sufficiently radioactive and have sufficiently long lives so that storage for long periods is necessary.

Were there to be reprocessing, according to the AKA report, it is "exceedingly important that the utilities will account for how the plutonium received from reprocessing will be utilized when they are concluding reprocessing contracts."[2]

The options available for utilizing such plutonium are:

To store it indefinitely for some undefined later use

To sell it for use in some other nation's reactor development program

To use it as fuel in Swedish breeder reactors

To use it as fuel in Swedish light-water reactors

To define it as a waste product and place it into a final repository along with other radioactive wastes.

These alternatives are illustrated in figure 5.2.

Radioactive Products Were There No Reprocessing

Were the spent fuel not to be reprocessed, the radioactive products would be retained in the spent fuel itself. The spent fuel would then be the ultimate waste product.

Reprocessing Is Not Necessary from the Waste Management Standpoint

Until recently, the nuclear industry simply assumed that the spent fuel would be reprocessed so that the plutonium could be recovered and used, for instance, as reactor fuel. With that assumption, there was no reason to think about anything else to do

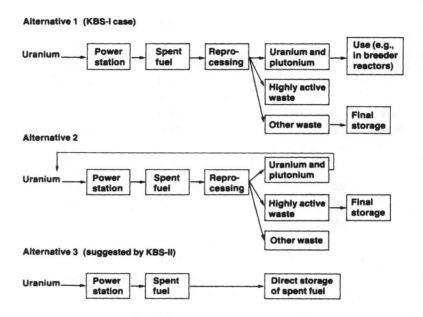

Fig. 5.2 Back end of the nuclear fuel cycle (existing alternatives).

with the spent fuel but reprocess it. All research and development for the management of the highly radioactive wastes assumed that the spent fuel would be reprocessed. There were no studies of treating the spent fuel itself as the ultimate waste product. To do so would have made no sense because it would have been uneconomic not to recover the plutonium.

A recent American study of the nuclear fuel cycle states:
Since reprocessing has always been assumed to be the next step in the fuel cycle, virtually no technical consideration has been given to longer-term secure storage of spent fuel or to the possibility of permanent disposal of spent fuel. The issues which must be considered include the feasibility of interim storage for periods up to several decades, the feasibility of disposal in geological formations as compared to resolidified high-level waste, and the long-term leachability

of spent fuel compared with treated high-level waste if the multiple barriers which isolate the waste are breached.[3]

The 1976 AKA study does not claim either that reprocessing was necessary nor that it was economic. The present American policy of indefinitely deferring reprocessing is evidence that American policy makers do not consider reprocessing necessary. The American policy has been sharply criticized, not with the argument that reprocessing is necessary for waste management reasons, but with the argument that it is necessary for economic reasons to recover the plutonium for use in breeder reactors. KBS-II claims that unreprocessed, spent fuel can be permanently stored with "absolute safety."

Some experts have suggested that reprocessing could somehow make the management of the wastes safer, even though the purpose of reprocessing is to recover the plutonium. A popular argument is that since plutonium is one of the most toxic and long-lasting (plutonium-239 has a half-life of about 24,000 years) components of the spent fuel, and as reprocessing would theoretically remove 99.5 percent of the plutonium, the wastes after reprocessing would not have to be stored for as long a period and therefore would be less hazardous.

Even if it had been demonstrated that 99.5 percent of the plutonium were removed during commercial reprocessing of high-burnup oxide fuel, there are a sufficient number of other highly active, long-lasting, alpha-emitting components of the waste so that this removal of plutonium would make little difference. As Keeny and associates concluded in 1977: 'Reprocessing and recycle of plutonium provide a way to reduce the long-term risks by reducing the amounts of transuranic elements in wastes. However, the magnitude of this effect is not large."[4]

Detailed analyses have shown no incentive for further removal of long-lasting components of the waste, that is, further "partitioning" of the wastes that would result from reprocessing (chap. 5, Disposal of Radioactive Waste Products).

The implications of isolating the plutonium must also be dealt with. To be sure, most of the plutonium would no longer be in the waste stream, but it would be somewhere. If the plutonium were recycled, that is, used as fuel in other reactors, then more radioactive waste would be created and the entire cycle would repeat itself. This recycling is, of course, the primary purpose for reprocessing and recovery of the plutonium. The plutonium could be recycled either in today's light-water reactors (there has been plutonium recycling on an experimental basis in the *Swedish* Oskarshamn reactors), or it could be used in breeder reactors.

Sweden has no breeder reactor program and no official plans for such a program. Therefore, were the plutonium recovered to be used for breeder reactors, Sweden would have to sell the plutonium to another nation.

Sweden would have the possibility of using the plutonium and the recovered uranium in her own light-water reactors. Such recycling could reduce the requirements for new uranium by up to 30 percent. However, were this done, the radioactive waste produced by these reactors would contain a much greater quantity of long-lasting transuranic materials. This could substantially increase the radiation exposures resulting from the KBS method of waste storage. The KBS report did not include an estimate of these effects but considered only the wastes that would result were Swedish reactors fueled with virgin uranium. Estimates of the radiation doses if a plutonium recycled core were used are included in the sensitivity analysis (chapter 14).

Whether the plutonium were stored, used in Swedish light-water reactors, or sold to another nation for its breeder program, the issue of safeguards arises whenever reprocessing is considered. This possible coupling between the civilian reactor industry and the proliferation of nuclear weapons is regarded by many as the most vexing political issue associated with the civilian nuclear program. There are many national and international programs dealing with one or another aspect of the weapons proliferation problem.

The quantity of radioactivity remains essentially unchanged whether or not there is reprocessing. Reprocessing is nothing but a process to separate some of the components of the spent fuel from one another.

It should also be noted that there is evidence that reprocessing makes the management of radioactive wastes more difficult. Reprocessing begins with all of the wastes contained in the spent fuel in a solid, rather insoluble form. After reprocessing, which is a combination of mechanical and chemical processes, wastes are contained in several forms each of which has to be treated in a separate way. The total volume of wastes is increased during reprocessing, although the volume of the solidified liquid component of the wastes from reprocessing is somewhat diminished compared with the volume of the spent fuel.

After reprocessing a permanent place to put the wastes still must be found.

Keeny and his colleagues concluded:

> The magnitude and complexity of future waste management problems depend in part on fuel cycle decisions. *Reprocessing would complicate waste management by broadening the spectrum and potential difficulty of problems.* . . . Without reprocessing and recycle, there would be time for a more orderly and assuredly more error-free process in both the management and disposal aspects of waste.[5] (emphasis added)

A Decision to Reprocess Does Not Preclude the Necessity of Final Storage Without Reprocessing

Even had Sweden now decided to reprocess spent fuel, there are several circumstances under which all or a portion of the spent fuel would not be reprocessed. These include:

> A later decision by the Swedish government to forego the potential economic value of the plutonium because of political considerations, presumably primarily implications for weapons proliferation

Refusal of the United States to permit the reprocessing of reactor fuel provided to Sweden by the United States

Inability of the French to reprocess at La Hague.

Similarly, a decision now not to permit reprocessing would not necessarily preclude reprocessing at some later time, if the spent fuel were stored in a retrievable manner, for example in a centrallager. Spent fuel placed in a final repository and encapsulated in the manner described in the KBS-II report is retrievable during the period prior to closure of the repository and is theoretically retrievable later, at least until there is failure of the capsule. The Stipulation Act offers the possibility of permitting loading without reprocessing.

As has already been noted, it has not yet been clearly established whether approval of an application under the Stipulation Act necessarily implies that reprocessing will be done. Should the finding be made, however, that the method proposed actually should be operationally meaningful, then it would appear that even were there a decision to reprocess, Sweden should be be prepared to store nonreprocessed fuel as well.

Disposal of Radioactive Waste Products

Management strategies for radioactive waste products must be suited to the chemical and physical form within which the radioactive isotopes are found. Radioactive wastes are found in various chemical forms and as gases, liquids, sludges, or solids. The nuclear properties are also important, particularly the radiological half-life. Current plans assume that radioactive wastes should be kept out of the biosphere for about twenty to thirty half-lives. Thus an isotope with a thirty-year half-life should be isolated for at least 600 to 900 years.

Simply stated, the basic problem with radioactive waste management is to put the waste into the most nearly stable, possible chemical and physical form, then put it where it is highly unlikely to be disturbed by natural processes or by man for twenty to thirty half-lives of the contained isotopes.

The radioactive wastes in the back end of the fuel cycle have half-lives ranging from very short periods to millions of years. Most of the troublesome fission products have half-lives of thirty years or less and so will require an isolation time of about 1,000 years. Many of the transuranic isotopes have half-lives of tens of thousands of years and so require an isolation time of hundreds of thousands of years.

These short and long half-life materials are mixed. It is theoretically possible to separate them using sophisticated chemical means, which are essentially very advanced forms of reprocessing called partitioning (further separation and recycling of the radioactive material of the wastes). Burkholder and his co-writers, among others, showed in a definitive 1976 paper that from a waste management standpoint there was no incentive for such partitioning. No commercial reprocessing currently under development includes partitioning.

Certain radioactive waste products are currently released into the environment at reprocessing plants. Foremost among these are hydrogen-3 (tritium), carbon-14, krypton-85, and iodine-129. As only very limited amounts of spent fuel have been reprocessed to date, relatively small amounts of these materials have been released and the resulting human exposures to radiation from these releases are well within the standards. It is widely recognized, however, that with the anticipated growth of nuclear power it will soon be necessary to trap carbon-14 and reasonable to trap other isotopes and to store them for appropriate periods.

PART III

Detailed Summary and Discussion of Remiss and Review Comments

6

How To Interpret "Has Shown"

The interpretation of "has shown" as it appears in the Stipulation Act is central to the development of the criteria to be used in assessing whether or not any given method satisfies the demands of the law. It is a matter of determining what are the scientific and technical requirements.

When can a technology be considered as "having been shown"? At what stage in the development process is it possible to make such a statement? A successful technical development process begins with a scientific hypothesis and proceeds through laboratory and field experiments, pilot plants, and then to routine application.

A technology generally could be considered developed when it has been tested without problems in full-scale operation under realistic conditions thereby showing that the technology does work, is financially justified, is environmentally acceptable, and is reliable. At the other extreme of the development process is the initial hypothesis. It is obvious that the technology described by KBS has passed the stage of initial scientific speculations, but it should be noted that it has only reached a very early stage in the development process. It must now be decided what stage of development is necessary to meet the demand of the Stipulation Act.

What demands should be placed on the validity of the evidence and the conclusions drawn from it? Is it necessary that there be general agreement in the technical and scientific communities that there is valid scientific documentation that clearly supports the conclusions reached? Is it sufficient that experimental data are presented which appear to support the claims made? Is it sufficient, when no experimental data are available, that several experts have a similar opinion? Is it sufficient to believe that the conclusions are correct?

In light of Parliament's demand for a very high level of confidence, it seems that a well-documented proposal with considerable unanimity behind it is indicated.

It is necessary to distinguish between the questions if a repository *ought to/shall have* certain qualities in order to be considered acceptable and if it has been shown that a repository *could* actually *be given* such qualities. This could be illustrated by the following sentence: It is assumed that the wastes are placed in sections with good rock with a water permeability that has been proved to be 1,000 to 100,000 times lower than the values used. This kind of argument must be kept apart from an actual demonstration that the desired qualities can be achieved, which is something quite different.

Some observers have attempted to make a parallel between the Stipulation Act in Sweden and the state of California's similar nuclear laws. The two situations are quite different, however, with regard to their waste management technology requirements. The California law clearly states that various technologies must "exist" and/or "have been demonstrated," and further that they have "been approved by the appropriate federal agency." The possible interpretations of these terms are fully discussed in reports from the California Energy Resources Conservation and Development Commission.[1] Further, phrases such as "have been demonstrated" can be defined through precedent in other American regulatory settings. Such is not the case for the "has been shown" that appears in the Swedish Stipulation Act.

There are several possible ways to proceed with the discussion of the standard to be used in finding whether or not the KBS method satisfies the Stipulation Act. At the onset, it must be decided whether the initial documentation of the method (contained in the KBS report and its background technical reports) is to stand alone or whether it is to be considered as though the various reservations and additions supplied during and since the remiss period are to be included as well.

It has been pointed out repeatedly that the KBS method is *not* a proposal in the sense of being a formal application to proceed with the construction and implementation of the method. Therefore, it cannot be evaluated in the same way as, for example, an application for construction of a nuclear power plant, a proposal where detailed engineering designs and specific safety analyses are available.

This dilemma is recognized by the KBS project staff and is mentioned in their report. The report is predicated on the assumption that it only need be shown that there is *a* method with analysis that indicates that it is possible and which shows that the general radiation exposure criteria that currently control the operation of nuclear power plants will not be exceeded.

A member of the Energy Commission's Review reference group has stated: "The Stipulation Act only asks for an *example* of how waste *can* be treated. . . . the actual handling may be quite another than what the KBS report says. The purpose of the law was, I think, to show the Swedish people that *at least one way exists*."[2]

There is also the related question of where the burden of proof lies. It appears to be necessary that the applicant, Vattenfall, prove that the demands of the law have been met and not that the various reviewers demonstrate that waste management *cannot* be done in a completely safe way.

Also the remiss process used for the Ringhals 3 application was not the usual one. That application consisted, in substance, of contracts for reprocessing services to be carried out in France and of the KBS-I report. The contracts were secret and therefore not available to most reviewers. The KBS reports had not been subjected to independent scientific review.

Although not explicitly discussed, the list of organizations included in the remiss reflected a recognition that what was going on was primarily a *technical review* and not the usual remiss. The task of the remiss organizations usually is, within the context of their own competence and interests, to comment on the desirability of proceeding with the described action. The strength of the remiss process is that it provides a formal mechanism for elements of society, holding very diverse opinions and values, to express their opinions as to whether a proposed action is acceptable, not whether it is technically possible.

When making a comparison, for example, of the lists of the remiss organizations selected for the review of the AKA investigation with those for the KBS reports, it is obvious that what was hoped for in the KBS case was a technical review, *not* a remiss. The Ringhals 3 application has not been subjected to a remiss in the usual sense of that process.

The remiss answers must be in the form of partial analysis. This is due, in part, to the limited time and resources that are available for the review. This means that assumptions must be made on conditions in other areas. When the assumptions that have been made in a specific area are compared with the results of the review, the assumptions are not always found to be justified. This means that statements of one reviewer regarding conditions in one area might be based on assumptions that have been shown to be invalid by other reviewers. This observation is important but is not intended as a criticism of the remiss organizations. It is a natural consequence of the treatment of complex systems. What is yet needed is to examine the remiss and review comments carefully to establish which parts of the KBS method have been subjected to critical review, what arguments have been drawn from those reviews, and what reservations and qualifications have been stated by the review groups.

It cannot be the responsibility of the technical reviewers to decide on the criteria that must be satisfied for a proposal to be satisfactory under the Stipulation Act. These criteria must evolve from the political process. In some cases, however, the reviewers have assumed either explicitly or implicitly a set of criteria and have drawn conclusions accordingly.

For the above-mentioned reasons this report does not include a listing of the general conclusions that have been stated by the various organizations and individuals who participated in the Swedish remiss or the other reviews. Instead, the subsequent sections of this report contain a step-by-step review of the KBS method and a detailed discussion of the technical and scientific evidence presented by KBS and the reviewers. Those factors that are important to the safety analyses have been emphasized.

In this report, the following factors from the reviews were analyzed:

that the source documents were published and reviewed

that experimental data are presented supporting the assumptions and conclusions

that those data that are presented are relevant

that all significant aspects of the problems have been considered.

Several of the reviewers have stated, within certain limits, that the level of proof implied by the above factors does not exist and, therefore, that their conclusions are qualified judgments.

7

What Constitutes "Highly Radioactive Waste Resulting from Reprocessing"?

The Stipulation Act does not, as noted, govern the management of all radioactive waste. There are entire classes of radioactive wastes that must be accounted for in the comprehensive Swedish waste management program but are not covered by the demands of the Stipulation Act. These categories of waste include those having little or nothing to do with the nuclear fuel cycle, for example, radioactive wastes from medical use, as well as wastes from some portions of the fuel cycle, for example, radioactive wastes resulting from uranium mining.

The primary difficulty is to define what constitutes "highly radioactive waste resulting from reprocessing." There is no dispute as to what wastes result from reprocessing. There are readily available, complete descriptions of such wastes (see chapter 5 for a brief review of the wastes from the back end of the fuel cycle).

The problem arises because the Stipulation Bill and its proposition are subject to different interpretations on this point.

To give only an example of the conflicting possible interpretations of the Stipulation Act, the following two passages are noted from the proposition:

> The highly active waste consists mainly of the so-called fission products that are converted to liquid form in the reprocessing process[1]

and

> These requirements imply that measures should be taken which, during all phases of the handling of the spent nuclear fuel, can ensure that there will be no damage to the ecological system.[2]

There are sundry waste streams from a reprocessing plant and other wastes that result from the use of recovered plutonium (see chapter 5).

The Swedish National Institute of Radiation Protection describes these different waste streams:[3]

> After reprocessing the following different paths for radioactive material must be dealt with:

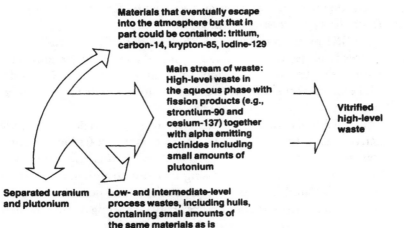

Materials that eventually escape into the atmosphere but that in part could be contained: tritium, carbon-14, krypton-85, iodine-129

Main stream of waste: High-level waste in the aqueous phase with fission products (e.g., strontium-90 and cesium-137) together with alpha emitting actinides including small amounts of plutonium

Vitrified high-level waste

Separated uranium and plutonium

Low- and intermediate-level process wastes, including hulls, containing small amounts of the same materials as is found in the high-level waste.

The KBS report includes discussion of these matters and concludes that only the highly active, liquid waste needs to be considered in the context of the Stipulation Act.[4]

Several of the remiss organizations and reviewers of the KBS reports commented on this issue.

FOA points out that the COGEMA contract does not clearly specify what the wastes will contain. FOA is also concerned about the gaseous wastes that would be released during reprocessing. FOA states, as did AKA, that it would be appropriate to include these volatile elements in the definition of highly radioactive wastes from reprocessing. They include in this category tritium, carbon-14, krypton-85, and iodine-129. FOA expresses particular concern about the possibility that carbon-14 would be released into the environment during reprocessing. If that were to be the case, the nordic collective-dose guideline would be exceeded by a factor of two or three. FOA goes on to indicate that methods are available to capture the carbon and that it should be possible to develop methods for its storage. Only through such capture and storage of carbon-14 could violation of the collective-dose standard be avoided. The FOA review includes a long and detailed appendix setting forth the problems posed by the release of iodine-129 and carbon-14.

Gothenburg University notes that other categories of wastes than the vitrified liquid waste discussed by KBS can be returned from the reprocessing plant.

SKI points out that the KBS definition of highly radioactive wastes from reprocessing, namely liquid wastes only, is consistent with previous Swedish usage, for example, in the AKA study. SKI expresses the opinion that such is appropriate for consideration under the Stipulation Act.[5]

PRAV also notes that the AKA investigation defines "highly active waste" to be the liquid wastes containing fission products that result from reprocessing. PRAV notes that the other categories of waste from reprocessing also contain alpha-emitters that must be isolated from the biosphere for a long period. However, these other wastes do not require artificial cooling during the first few years after removal from the spent fuel.

SSI begins its discussion of the problem of defining what is highly radioactive waste from reprocessing with a detailed review of the statements included in the AKA reports. SSI then states that it interprets the point of the Stipulation Act to be to guarantee that a safe method exists for dealing with the radioactive waste that has the greatest potential hazard, namely the high-level liquid waste. SSI points out that this does not imply that it considers the management of other categories of radioactive waste to be free of problems. SSI ponders, as did FOA, whether the gaseous products, such as tritium, carbon-14, krypton-85, and iodine-129, should be included as high-level wastes. SSI notes that these are now released into the atmosphere.

> Though not formally demanded in the Stipulation Act the Swedish National Institute of Radiation Protection has considered it reasonable to make a complete review as to the most critical radioactive materials in the waste. If the purpose is to show that no part of the radioactive materials can lead to unacceptable exposures of radiation it might be considered too narrow an approach to stick to narrow technical definitions.[6]

SSI, however, regards the demands of the Stipulation Act for a safe final deposition fulfilled, thereby assuming the leakage from the repository as presented by KBS to be correct.

Rydberg and Winchester's review for the Energy Commission includes detailed comments under the heading: "What is high active waste?" They note that the definition used by KBS is consistent with the formal definition as used by a variety of authorities. They conclude: "The KBS definition of 'high-active waste' is in accordance with definitions used by most legal and official organizations. It can also be considered to agree with what is accepted within the trade."[7]

These reviewers, however, then suggest that in their view, based on a consideration of the radioactivity contained:

(i)   The liquid raffinate from the first extraction cycle (and streams added to it)

(ii)   the cladding hulls

(iii) the gases released in the first extraction cycle.

Only the first (i) is treated by the KBS report, and only in the form of solidified and vitrified material in stainless steel canisters.

This is of particular importance because the technology for handling the other wastes is insufficiently developed, although the problem is of much lesser magnitude than for the high-level liquid waste. The KBS report does not indicate the expected fate of the cladding hulls or of released gases. Although this may not be required by the Stipulation Law we are aware that these other waste categories may be returned to Sweden. Consequently we think that these other wastes must be considered within the general frame of high-active waste problems.[8]

8

What Is "Absolute Safety"?

A Framework for Analysis

The Stipulation Act calls for the reactor owner to show how and where the final storage of the highly radioactive waste resulting from reprocessing can be effected with "absolute safety." In the proposed bill the meaning of "*absolute safety*" was somewhat elaborated on. The committee of Parliament preparing the decision has made some further comments regarding the interpretation of the concept.

> As was made clear in the explanatory statement of the bill (page 24) "very strong safety requirements" are intended where "the basis must be that the high active waste from reprocessing and the spent fuel that has not been reprocessed must be separated from all life." A number of criteria

for the safety judgment are stated in the bill. It is underlined that the storage must meet the requirements from a radiation protection point of view aiming at protection against radiation damage. Furthermore it is pointed out that the waste or the spent nuclear fuel is isolated as long a time as is required for the activity to diminish to a harmless level. The risk of release to the biosphere is also recalled. The Parliament Committee finds "absolute safety" to be an adequate formulation of the very high level of safety that obviously is required. That a clearly "draconian" interpretation of the safety requirements is not intended, is expressed in the recently referred quotation from the explanatory statement.[1]

Draconian was not defined, and this concept also needs a working definition.

It appears that the criteria for absolute safety must await an action by the Swedish government.

It is explained in the bill that the final storage primarily "has to meet the requirements of radiation protection aiming at protection against radiation damage. The repository must be arranged so that the waste or the spent nuclear fuel is isolated as long a time as is required for the activity to diminish to a harmless level."[2]

The Swedish National Institute of Radiation Protection has discussed the aspects of radiation protection. The Institute follows the recommendations of ICRP. New instructions for the limitation of releases from nuclear power stations were issued in 1977. The releases permitted by the 1977 regulations, according to the Institute, should be regarded as absolutely safe in terms of the Stipulation Act. The Institute further assumes that waste management is absolutely safe if it does not lead to consequences worse than those that might result from other portions of the fuel cycle.

The harmful effects from ionizing radiation have been recognized for many years. National and international standards and recommended guidelines intended to limit human exposures to radiation have been developed. A basic feature of these standards is the assumption that there is damage from all exposure regardless of the amount. The damage is assumed to be directly proportional to the size of the exposure received by an individual or a population.

The Swedish Radiation Protection Agency has also adopted the guideline of a maximum dose rate to members of the public of 10 millirems per person per year from the operation of a nuclear power program.[3] The maximum future annual radiation dose to any individual in a critical group should not exceed 50 mrem and preferably should not be higher than about 10 mrem to meet the requirements that are now stipulated for radiation doses around nuclear power plants.

The Swedish National Institute of Radiation Protection has adopted the general guideline of a maximum cumulative dose commitment of 1.0 manrem/MW(e)-year for the entire fuel cycle.[4] Half of that, or 0.5 manrem/MW-yr, has been allocated to the reactor itself. The remaining portion must be allocated to all remaining steps in the fuel cycle. Waste management is but one of these other steps. When the 0.5 manrem/MW-yr is allocated among the remaining parts of the fuel cycle, it is likely that only a minor portion can be alloted to the final deposition of radioactive wastes. This portion will be less than 0.5 manrem/MW(e)-yr as experience from reprocessing already done indicates that the reprocessing exposure runs relatively large.

It is worth noting that the adoption and shaping of regulations for radiation protection have become more restrictive as time goes on. This has been due to two factors: changing public values have emphasized increased safety, and new scientific evidence has shown that the harmful effects associated with ionizing radiation were greater than had earlier been thought. Both of these factors are still operating. Experience strongly suggests that the trend toward lower permissible levels of radioactivity could well continue.

There are no specific standards as yet for radioactive waste management. It is not uncommon for the Swedish Nuclear Inspectorate to adopt American standards and guidelines for other phases of the nuclear fuel cycle, for example, reactor safety.

Standards and criteria for radioactive waste management, including criteria for the location of waste storage sites, are currently under development in the United States. In some cases these standards and criteria have already been published in draft form.

If the regulations for waste management to come follow the same pattern as the present standards for radiation at nuclear power plants, economic and social factors will also be central to the management methods. The Swedish reactor regulations include as a general regulation:

> §9. The release of radioactive substances from nuclear power stations *shall be limited to the extent that is reasonably achievable, taking into account the economic and social consequences of every measure for the limitation of releases* and risk that the radiation dose to the personnel may increase when releases into the environment are reduced.[5] (emphasis added)

This implies, as has been demonstrated in practice, that the economic cost of any given measure to reduce environmental releases must be known and compared with the benefits derived were the releases reduced. While not including the value as part of its recommendations, the Swedish National Institute of Radiation Protection calls attention to the U.S. Nuclear Regulatory Commission's recommended value of US $1,000 as justifiable expense to reduce the dose commitment by one man-rem.[6] The KBS report, as well as some of the reviews, includes estimates of the dose commitments that may result were the KBS method implemented. There is, however, no economic analysis of the method.

According to SSI, the application of the concept "absolutely safe" will be a combination of the reliability of the dose estimates and the estimated dose level in relation to the basis for forming a judgment.

The Swedish National Institute of Radiation Protection has in another connection stated:

> As the methods for final deposition of the high active waste which have been discussed, for instance, dumping in salt-

mines or in primitive rocks, are intended to separate the waste from all life forever, the collective dose commitment from the waste dumping of "normal operation" is zero. The problem will instead be to try to judge what events might cause the radioactive materials to be released in spite of all calculations.[7]

There are two distinctly different steps necessary for an assessment of the risks of radiation exposure. The first, emphasized by several of the reviewers, is the estimation of the quantity of radioactivity to which humans might be exposed as a result of the waste management method. The second, addressed only by SSI, is an estimation of the quantitative relationship between radiation exposure on one hand and the observed human damage on the other, for instance, life shortening, cancers, and genetic damage. SSI points out that there is a large uncertainty in the estimates of biological damage resulting from radiation. The risk *might be* much smaller, or even absent, at very low dose rates although this is not now considered very likely. The risk also *might be* larger, but hardly much larger, as it by then should have become apparent in the many epidemiologic studies that have been carried out. SSI considers that a factor of two higher is not unlikely although some researchers now think, on the basis of new epidemiological studies, that the risk might be five to ten times larger. For present purposes SSI thinks that these differences in the various risk estimates are rather unimportant.

In a *DsI* report B. Persson discusses the plutonium criteria. His conclusion is that plutonium limits must be made more stringent by a factor of five to ten to achieve the same level of protection that is achieved with present criteria for radium.[8]

The radiation standards act on the dose absorbed by an individual. To make the calculations one has to know the intake of various radionuclides. The intake is influenced by the concentrations of various nuclides in food, water, and air. Some of the absorbed amount leaves the body after a certain time. To judge the connection between a certain inflow of radioactive substances in the biosphere and existing radiation standards, all the factors that affect the movements of the nuclides must be considered.

According to the Safety and Environmental Expert Group of the Energy Commission, new data have shown that under certain circumstances the intake of plutonium, neptunium, and americium at low concentrations from food might be several hundred times larger than what has been observed in older experiments.[9] This might lead to a reduction of the limits for these elements in drinking water by a factor of 100 to 1,000. This would obviously affect the present risk estimates with a similar factor, especially since the dose from neptunium already dominates the picture as given by KBS.

Different organisms are susceptible to ionizing radiation in different ways. If an organism in the ecosystem is seriously damaged, this might lead to considerable changes in the whole ecosystem. Such problems have not been discussed in connection with the KBS review.

Considering uncertainties of this kind and the limited information available on many important factors affecting the routes of the transuranic materials in nature, the questions arise concerning taking the long-term character of waste management into account. Should present limits be provided with an extra safety margin to cover the uncertainties derived from limited knowledge of the effects of a final deposition?

One possible way of dealing with these uncertainties would be to ask for a large gap between calculated "worst cases" based on present knowledge and doses that might be acceptable.

The size of the gap would have to be related to the degree of uncertainty. As the latter is not well known in many instances, an estimation is difficult. The judgment would therefore have to be done in the political process.

Time Horizon

The time horizon of the present discussion is new for the societal decision-making process. Society generally deals with decisions with implications for periods anywhere from a few to a hundred years, with implicit implications at times for much longer spans. An example of this is the effect of carbon dioxide emissions on climate, an issue with very long-term implications and "irreversible" features similar to those for radioactive waste. The present issue, however, forces an explicit treatment of the long period of thousands to millions of years.

The issue can be subdivided into three parts: one technical, one administrative, and one dealing with values.

Can we make reliable predictions today on how different technical designs will behave thousands of years from now? Perhaps, but the issue has to be addressed and opinions differ. It is not necessarily sufficient that a majority of experts hold the same opinions, as we are not looking for the most probable answer that is interesting but for the correct one that is unknown. This lack of knowledge raises the question of how the decision process should take the uncertainties in the records into account.

The administrative problem consists of creating conditions such that the technical solutions proposed and accepted as safe actually will be carried out.

Turning to the question of values: What is an acceptable release of radioactivity from a final repository in the distant future? Do we have to consider man on earth in a time perspective of up to one million years? Should they be valued as if the doses of released radioactivity occurred within a few generations?

This is a nontechnical issue, and it has to be recognized that it is an ethical and political question that should be handled explicitly by the political process and not implicitly by scientists, that it has to be based on information that is as complete as possible about what can justifiably be said from a scientific perspective.

There seem to be technical advantages in postponing the final disposal of waste a couple of decades. This means that a design for the repository does not have to be ready today. However, the Stipulation Act asks that it be shown there is at least one solution before waste is generated, as it may be very troublesome to solve the problem later.

SSI considers it is not totally unreasonable to consider the radiation damage from the first 10,000 years as is done by KBS. This span is of the same size as man's entire cultural history and also the same as the intervals between ice ages. It is more difficult to see how the collective doses that span periods of up to millions of years or more could be compared more meaningfully with collective doses that fully affect the next generations.

SSI maintains that the judgment of damage in a future more distant than tens of thousands of years is not a technical and scientific question suitable for an expert agency using today's standards for radiation doses. This is a political-ethical question to be handled by responsible politicians.

Scientists at the University of Gothenburg argue that a decision regarding radioactive wastes must be balanced between the risks and the benefits of the activity. If the loading of Ringhals 3 would lead to water spilling past hydropower stations, the risk from nuclear waste would obviously be unacceptable. But if the alternative would be a coal- or oil-fired power station, the judgment is much more difficult, and sufficient material is not available. It should be remembered that a great deal of the responsibility and the costs for waste management burden others than those who benefit.

Experts at the University of Lund argue in light of the very long times involved that special care is required as the steps taken cannot be reversed.

They further raise certain moral aspects:

> In today's society it has become evident that humans have a right to make use of nature and its resources, to a great extent. Not until lately have we realized that nature's resources are limited and that we are part of ecological cycles into which we—assuming we want to survive—must play our roles. Particularly during the last century demands for consideration to shepherding nature's resources have marked our views of society's future. Demands have been put forward for a continued accelerated development and for moral responsibility also to future generations.
>
> This revised look on nature's resources has not only scientific and political aspects, but also ideological and moral ones. The moral aspect comes forward when asking how the present generation by its actions determines the situation on earth for future generations. By keeping the radioactive

waste, for the first time we carry out an action, conscientiously, the results of which will follow after thousands of years perhaps.

Therefore, the problem should not only be subject to technical security assessments but also be looked upon from a moral point of view. Irrespective of how secure the present development may seem, one must nevertheless ask, with such a long time schedule in mind, if uncertainties still must not prevail. It is, therefore, a justified demand that the question of waste being finally disposed of should be subject to discussion wherein these aspects are considered.

Points of View on the Issue of Absolute Safety

It should be noted that a number of very different approaches to dealing with the absolute safety issue has been presented.

KTH argues that the concept of absolutely safe puts a scientific group in a difficult position, as it is obvious that it is not possible to make a determination having scientific validity that something is absolutely safe. KTH finds the concept uninterpretable, therefore it cannot take a stand on whether the KBS plan fulfills the law or not.

KTH and CTH point to international radiation-protection work and the desire to keep doses from nuclear power well below that of natural background radiation.

One reviewer, Professor Rydberg, argues:

Abundant and cheap energy has been a major cause in relieving the Swedish people from freezing and starving, and from diseases and crippling hard labor. . . . we will run out of these fossil energy resources in a few generations. . . . the so-called continual energy sources like waterfalls, wind, and sun will be insufficient. . . . Thus, faced with the necessity of using nuclear energy we are also faced with its problems . . . (then after an evaluation of the KBS method, particularly the adequacy of the titanium-lead capsule). . . . When the lead canning dissolves, it may influence the retention conditions for the waste nuclides, either causing increased retention or decreased retention. This leads to an uncertainty about ground conditions which could be an obstacle for making predictions with the geochemical model . . . if the lead shielding is not used or is replaced by another shielding material. . . .[10]

In Rydberg's view, the KBS method is uncertain, but if the KBS method is inadequate, a better suggestion will turn up instead; as nuclear energy is a necessity we simply must face its problems and solve them.

Another reviewer (CTH) also suggests that "The great number of experts that have been engaged is a guarantee that the problems have received an all-round and objective examination. . . ."

This statement can be compared with Professor Hillert's, KTH, quoted below. Also it neglects the fact that, as pointed out by SSI, what are acceptable exposures to furture generations is an ethical and political question and not an expert question.

As has already been noted, some reviewers regard "absolutely safe" to be operationally defined as meeting the appropriate regulatory standards, that is, those of SKI and SSI. One difficulty with this approach is that there are no standards for waste management, and in the setting of standards for nuclear power reactors, it was not necessary to address directly the problem of large radiation effects that occur far in the future: the situation where one generation enjoys the benefits and future generations bear the costs. This situation is now acute given that it has been shown that the KBS method would likely lead to future collective dose commitments significantly in excess of those postulated for reactors (a portion of the fuel cycle for which SSI has issued regulations), even if the dose to an individual or a generation might be small.

During the Energy Commission review of the KBS proposal, a member of the reference group, Lars Norberg, brought forth quite a different interpretation:

The basic aim of the Stipulation Act is obviously a moral one: what commitments of risks are morally acceptable to leave to future generations, especially under the following conditions: (a) when the present generation takes all the benefit and future generations gain no benefit, but most of the risks, and (b) when the future risks cannot clearly be defined and estimated?[11]

As can be seen, radiation protection criteria of today deal with a situation where a cost-benefit analysis can be done, but the benefits in this case are given to one generation and the risks to another.

Cochran, commenting on proposed American waste management criteria, has discussed this problem and suggests that:

> The cumulative risk to all future generations from radioactive waste should be less than, or (considering the uncertainties in the calculation) comparable to, the cumulative risk to all future generations from the original uranium resources from which the radioactive wastes were derived, assuming these uranium resources were unmined.[12]

A criterion of this type is implicitly suggested by KBS[13], EKA[14] and Rydberg[15] when including uranium ore in figures describing the relative danger of the wastes. This criterion presumes that all wastes from the nuclear power cycle are included, not just reactor wastes.

The discussions of benefits and costs allocated to different generations are not clear-cut. It could be argued that the conditions for future generations are also affected by the use of nuclear power in the sense that carbon dioxide, for instance, released from combustion of fossil fuels is reduced, nonrenewable resources are consumed at a slower rate, and so forth. However, the balancing of these benefits is not possible at present; the benefits can at best be qualitatively described. The criteria suggested by Cochran have the advantage of not exposing future generations to any unjust leftovers from our activities, if not improving their situation. Furthermore, the criteria are quantitative and thus can be applied. Unfortunately, this has not yet been worked on.

Cochran goes on to argue that other suggested criteria suffer from serious deficiencies. For example, it has been suggested that the risks from waste disposal should be small compared with natural background radiation. This is based on a comparison of a cost with an unrelated cost. Another suggested criterion is that the risk of waste disposal should be small compared with other risks in the nuclear fuel cycle. This is a benefit-cost comparison where the benefits are the same for all components. It favors those that benefit from the use of nuclear power at the expense of future generations that do not share in these benefits, according to Cochran.

The same distribution problems arise between those who enjoy the benefits in one location and those who bear the costs in another (for instance, global exposure to iodine-129 or carbon-14 or release of sulphur from oil burning).

How To Handle the Concept of Absolute Safety

The analysis may be carried out in three relatively distinct steps:

1. Is the waste management proposal contained in the KBS report scientifically sound? This is dealt with in the discussion in Parts III and IV below, based on the review comments and additional work with the KBS models.

2. Is the waste management proposal in the KBS report complete? This will be discussed in Part III regarding technical and scientific points brought up in the review process.

3. Are the human exposures to radioactivity that appear possible, were the proposal implemented, acceptable? This is a political judgment based on records worked out by authorities and other experts where discussions about the above possible criteria must be compared with the exposure risks as discussed in parts III and IV below.

The Approach Taken by KBS

KBS was required to show that something is absolutely safe and had to choose among different approaches. In Wynne-Edwards's description KBS has chosen the following method:

1. Define the problem precisely and narrowly
2. Subdivide and clarify its parts
3. Analyze these parts and determine the factors involved
4. Design a model for each part of the problem
5. Assess the factors within these models and assign probabilities
6. Conduct preliminary structural design to satisfy the assumptions developed above
7. Assemble the parts of the proposed solution.

Wynne-Edwards states that the approach taken by KBS has been standard engineering and technical practice.

The radionuclides have to pass a number of barriers to reach the biosphere. These include the glass in which the radionuclides are incorporated; the encapsulation of the glass in a three-layer, metallic capsule; and the deposition of the capsule at a depth of 500 meters in good rock. It is argued that any one of these barriers gives protection against dispersion of the waste.

The properties of these barriers are discussed below. If intact, a barrier provides complete protection from the waste. If damaged the barrier may still delay the influx of the nuclides into the biosphere long enough to reduce their significance due to radioactive decay.

In estimating the effects of the final storage expressed as radiation doses to man, mathematical models have been used. There are three models: ORIGEN, calculating the amount of various nuclides in the waste; GETOUT, calculating the amount leached out of the glass and transported through the rock after breakage of the capsule; and BIOPATH, calculating the movements of the nuclides in the biosphere. The models are coupled together as shown in figure 8.1.

Values inserted in these models are claimed to be conservative. This approach has been chosen by KBS to allow certain margins for the present lack of knowledge and uncertainties in the variables. If it is clear that these assumptions are well on the

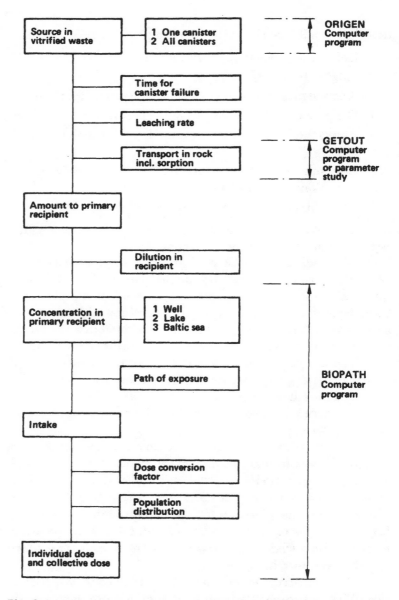

Fig. 8.1 Radiation doses from radioactive elements released from final repository (schematic illustration of calculations). (Source: KBS-I, IV, 85, fig. 6.5.)

safe side, KBS then argued that the result will be a high estimate of the radiation dose.

The approach using conservative numbers instead of the more common approach—best estimates of the value of each parameter together with an analysis of the effects of uncertainty—has been criticized by many reviewers. Rydberg, CTH, and researchers at the University of Uppsala are of similar opinion. The later method would have facilitated a more stringent analysis. Also less likely events could then be assigned a probability value.

Wynne-Edwards considers that the steps taken by KBS are

essential components of the whole study, but if this alone is done, the result may be a fragmented analysis of the problem in which the essential synthesis or integration of the system is downplayed or omitted.

To avoid the dangers of a piecemeal analysis and to place it within not only the technical, but also the social and environmental context of the whole system, answers to the following general points are all necessary:

1. Are all the relevant facts known and understood?
2. Are all the external effects known?
3. Are there satisfactory internal control and feedback mechanisms to regulate the system?
4. What are the costs and who will bear them?

Many questions may be raised under points 3 and 4. Wynne-Edwards considers it "essential, however, that they be recognized as having equal, if not greater, validity and importance to the points already answered. Otherwise there is a danger that only factors susceptible to an engineering solution will be dealt with."

The California Energy Commission report describes the KBS approach as "both an excellent summary of the state of worldwide knowledge on nuclear waste disposal issues and a sound formulation of a consistent methodology for dealing with the back end of the nuclear fuel cycle."

However, the commission believes that the work is

constrained by a certain lack of fundamental scientific knowledge in the application of the earth sciences to the problem. As a result, current plans and engineering regimes *require a departure from the scientific method and substitute engineering hypothesis and belief for scientific understanding. This engineering approximation requires scientific validation.* The concept of demonstrating how and where the solution to the waste disposal problem will take place requires conducting experiments to verify the engineering models used to support a level of confidence that the waste system will work as planned. As well, hypothetical geologic attributes must be found in nature and tested to confirm that the associated values of the geologic media exist in a real location. The KBS plan itself admits to these shortcomings by providing a reasonably long period of time (at least 30 years) between the receipt of waste in Sweden and final action to commit such wastes to an ultimate repository.

We would, however, point out several implications of rigorously applying the scientific method to this type of technical, social and political issue:

Inherent in the application of the scientific method is that the hypothesis may simply be proven wrong. Validation by experiment is necessary to determine whether the postulated concept will in fact work.

The scientific method does not always take into account the temporal necessity to make decisions in a manner which recognizes the economic, social and political dimensions of the issue (emphasis added).

A similar concern has been expressed by Professor Hillert:

The KBS plan is a very complex technological problem and it involves an enormous number of risks. A very large number of experts have been engaged in order to reduce and estimate such risks. In an undertaking of this size it is possible that a gradual screening of risks occurs.

Each expert will probably consider a number of risks and it is natural that he discusses only some of them in his written document. It is to be expected that he leaves out risks which he considers being less probable. This could be unfortunate because a risk which has a small probability could be very important if its consequences are large and it may often be difficult for an individual expert to realize what the consequences could be. Furthermore, it is natural that the expert leaves out risks which are difficult to estimate and concentrates on such risks which can be expressed in numbers. Finally, he may not consider the consequences of uncertainties and limitations in the definition of the problem he has been asked to treat. As the result of the individual expert is being carried through larger groups of experts and finally included in the main report, it is natural that more of the risks are omitted from being discussed or even mentioned. In order to evaluate the reliability of a report from such a large undertaking as the KBS project it is necessary for the referees to examine to what degree important risks have been omitted due to such reasons.[16]

One must bear these factors in mind when making an overall judgment of KBS's suggestion.

9

Central Spent Fuel Storage Pool (Centrallager)

The spent fuel elements have to be stored for some time after being removed from the reactor as reprocessing will not be available until the mid-1980s. For the first years of reactor operations, storage is available at the stations, but for longer storage, this space is too limited. The centrallager is intended for this storage.

The centrallager is designed for storage of spent fuel elements in pools with water for cooling. The pools are placed in subsurface rock at a depth of 30 meters.

KTH, SKI, SSI, Atomenergi, Winchester, and PRAV have discussed the centrallager. Atomenergi and Winchester consider the treatment of cooling system failure to be insufficient. Even if more difficult cooling problems have been solved at power stations, Atomenergi stresses these problems must not be overlooked in waste management designs.

PRAV underlines the importance of satisfactory sabotage protection, which is simplified by the underground placement.

According to KTH, it is doubtful whether 30 meters of rock would be sufficient protection in case of war. KTH questions, against this background, if the central storage would become a final storage if the roof came down on the pools?

The centrallager is now subject to review under the Atomic Energy Act. The centrallager is the only part of the KBS method that has actually been proposed for implementation. A formal request for permission to construct the centrallager was submitted by SKBF in 1977. SKI will propose the instructions needed for the detailed design. SKI and SSI will then make a more detailed review under the Atomic Energy Act.

10

Transportation

The KBS plan assumes transport of spent fuel to the reprocessing plant in La Hague, France, and of vitrified high-level waste from the vitrification plant back to Sweden. These shipments are intended to go by ship with additional land transportation as required.

SKI stresses that routine transport of spent fuel occurs internationally; for example, more than 250 shipments have been made to La Hague without any mishaps.

SSI notes that the transportation activities will be large and include two or three cargoes per month. In a longer time perspective the risk of mishaps and accident cannot be neglected. Even if doses resulting from such individual events were to be much smaller than from routine operation of power stations, it is important to devote considerable attention to radiation protection activities should the KBS plan be implemented. According to SSI, there is no evidence that radiation protection problems would become an obstacle for the implementation of the KBS plan.

The importance of preventing the radioisotopes from escaping the capsule is demonstrated by SSI in a calculation that shows that one accident where *all* the important nuclides would be released into the Baltic would increase collective doses by a factor of a hundred compared with the fallout from weapons testing. A collective dose on the order of 10^5 manrem/MW(e)-yr would result. Considering the probabilities for accidents and the probability that neither the capsule would hold nor the capsule be retrieved, the SSI finds it unlikely that the required low values for these numbers would not be reached.

According to PRAV, type B packing would be needed for transport of spent fuel and high-level waste. KBS regards the IAEA recommendations as requirements. These recommendations include: free-falling 9 meters against a rigid object, free-falling 1 meter against a massive (15 centimeter diameter) steel cylinder, heating to 800°C for thirty minutes, and submersion in 15 meters of water for eight hours.

PRAV, University of Lund, and Winchester each notes that the submersion test did not include the conditions to which the container would be subjected in an accident where both ship and container sink.

The IAEA panel considers the transportation risks acceptably low if IAEA recommendations were followed. They judge that experience to date has been reassuring. The assumptions made by KBS are considered conservative; for instance, the probability that transport casks might be damaged due to a collision at sea had been put at 0.1 by KBS whereas the panel believes something between 10^{-3} and 10^{-4} to be a more probable value. The principal risk, according to IAEA, "seems to be associated with the possible loss of a cask into deep water after a collision. However, national and international assessments have shown that the radiological consequences of such an event would be small."

STU concludes, after having pointed out several technical points for improvements in the handling scheme, that the conventional technique proposed should work satisfactorily.

KTH notes that the risks of sabotage must be considered.

The KBS estimates of accident probability at sea were based on global, average statistics. As pointed out by Scandpower and University of Gothenburg experts, the heavily trafficked routes through the Sounds and the British Channel would motivate a higher than average accident frequency, but the special precautions that would be expected for nuclear transports might motivate a lower frequency. University of Gothenburg scientists, using KBS numbers, calculate a 5 percent probability per year for an accident, which means that it "is as likely as not that an accident occurs before year 2000 (p. 2)."

KBS, supported by PRAV, claims that the releases of radioactivity after an accident would be very small. University of Gothenburg (UG) points out that this is based on the assumption that the capsulation will stay intact and that the capsules will be recovered quickly. UG doubts that quick recovery can be assumed considering the water depths and the technical difficulties involved. Sjöfartsverket, in a letter to PRAV, asked, however, that nuclear transports be limited to areas where ships can be recovered with an accessible and established technology. PRAV concludes such technology exists, citing as examples the recovery of nuclear submarines and of drums of lead compounds.

The California Energy Commission comments that in a report of the Lawrence Livermore Laboratory, funded by the U.S. Nuclear Regulatory Commission, transportation is regarded as the most vulnerable part of the back end of the fuel cycle and is calculated to outweigh by far all other risks entailed in waste management.

11

Reprocessing and Vitrification

Vattenfall, in its proposal to load the Ringhals reactor under the Stipulation Act, chose the alternative based on reprocessing of the spent fuel before final storage of the waste. Reprocessing is intended to be done in France, and contracts regarding this service have been signed between SKBF and the French company COGEMA.

A major issue in the context of the Stipulation Act is whether reprocessing of the Swedish spent fuel would actually take place. The validity of the contracts and the legal situation reflected by the contracts have been discussed previously. This section includes brief comments on the technical situation with respect to reprocessing technology.

Reprocessing technology was not extensively discussed in the review process. SKI, SSI, and PRAV made some comments either in conjunction with the KBS review or in the separate review of the 620-ton contract for the yet to be built French UP3-A plant.

SKI considers the COGEMA experience with reprocessing to be solid, that this should guarantee no technical problems arise in the construction of the plant, and that the plant will be taken into operation as planned. SSI gives no reason to question that reprocessing could occur in a radiologically acceptable way.

The Expert Group for Safety and Environment of the Energy Commission (EKA) has reviewed the state of the art of reprocessing. Reprocessing has taken place, as part of weapons programs, since 1944. Low-burnup spent fuel from gas-graphite reactors has been reprocessed in both military and early civilian reprocessing plants. Higher-burnup (measured in MW-days) means higher content of radioactive material per ton of fuel. The transition to higher levels of burnup, as in the oxide fuel from LWR, was, according to EKA, more difficult than anticipated, and reprocessing technology requires additional time to achieve the degree of industrial maturity of the reactor program.

LWR oxide fuel has been reprocessed in West Valley, New York, Winscale in the United Kingdom, and in the demonstration plant at Eurochemic in Belgium. Presently, LWR fuel is being processed in the German WAK pilot plant, at La Hague, and in the new Japanese Tokai Mura plant. In the latter plant, two batches of four tons each have now been processed.

West Valley was operated between 1966 and 1972. A total of 624 tons was processed, 62 tons of which had a burnup level over 15,000 MWd/ton (about half the level of LWR oxide fuel). The plant was shut down, in part, to improve radiation protection for employees. While those improvements were being made, new safety regulations were issued by the NRC. It was considered uneconomical to meet these regulations. The plant was therefore permanently closed.

The present Winscale plant has been in operation since 1964. No oxide fuel has been handled since 1973 when an accident occurred in the section for oxide decapsulation and dissolution. A total of 19,000 tons of spent fuel has been processed in Winscale, 100 tons being LWR oxide fuel.

Eurochemic was operated between 1966 and 1974. The plant was owned by thirteen nations, including Sweden. About 200 tons of oxide fuel were processed. The operation was closed after the German and French decision to concentrate on their own reprocessing.

The La Hague plant started operation in 1967. The capacity was 800 tons of low-burnup fuel per year (of about 5,000 MWd/ton). To handle oxide fuel, a new head end was installed with planned capacity of 400 tons per year by 1976–77, increasing to 800 tons per year in 1981. To date two batches of LWR oxide fuel have been processed. The first (1976) included 15 tons and the second (1978) 54 tons. The operations are reported to have been compatible with existing radiation protection standards. Reported values from other plants indicate, with the limited experience available, that individual doses exceed the standards only in exceptional cases. The collective doses vary between 0.1 and 1.9 manrem/MW(e)-yr, to be compared with the recommendation for the *total* nuclear fuel cycle of 1 manrem/MW(e)-yr.

The coming situation where fuel with some five times more radioactivity per ton than before must be handled on an industrial scale implies a situation that differs from the experience to date. The availability of the plants, and thus their total output, therefore remains questionable. Additional uncertainty is added by the scientific debate of the radiation protection standards for elements such as plutonium. More stringent standards would add to the technical difficulties of maintaining operation.

12

Intermediate Storage of Vitrified Waste (Mellanlager)

The waste cylinders returned from France would be stored in the mellanlager for thirty years. After that they will be put into the lead-titanium capsule and transferred to final storage. The mellanlager is proposed to be located with a 30 m cover of rock. One reason for the thirty-year storage time is to let the heat generation from radioactive decay decrease from 1200 Watt to 525 Watt per cylinder. Another reason is to postpone the time when encapsulation and final storage are due in order further to develop and optimize these steps in the handling scheme.

According to SSI, there is no evidence that it should not be possible to handle radioactive materials in the mellanlager in accordance with instructions the Institute may issue.

STU observes that the technology to be used is conventional and should present no problems.

Winchester does not think the first reason for the mellanlager is sufficient to justify the facility since a factor of two heat reduction per canister could be achieved by making each canister initially contain half as much radioactivity. Therefore, he assumes the primary reason is to provide additional time to design the final disposal plant, locate a site, and so on. If this time is necessary to assure safety in waste disposal, this implies present knowledge is not sufficient.

The facility presents a target for terrorist activities. To prevent these from occurring, an adequate protection plan should be presented.

SKI finds the proposed cooling system based on forced ventilation to be adequate. In case of fan failure, natural convection would be sufficient to keep the canisters at a temperature of 340°C, well below the 550°C recrystallization temperature. Winchester, however, envisages situations that might arise where the ventilation shafts are blocked and the canisters, therefore, would overheat and would have to be reprocessed in some way. Therefore, there is a technical risk that extends over sixty years in having to maintain, by human attention, the ventilation system of the intermediate storage facility.

SSI also notes that the possibility of reducing the need for intermediate storage through a reduction of the amount of waste per canister must be considered. Otherwise it might be that the construction and starting up of the final repository would take place after the operation of the reactors had ceased, and thus there might be no economic support from the electricity production company to cover the costs involved. A long period between the operation of the reactors and final disposal seems unsatisfactory. From a radiation-protection point of view, late construction of the facility would permit optimal incorporation of research and development results, but this is only valid if economic and institutional guarantees are created so that the repository eventually will be built.

Economic and institutional measures of this kind have not been discussed by KBS. As such steps taken now are of a social character and might be upset over the next decades by events that are difficult to predict, it would be of some interest to perform a detailed safety analysis of the intermediate storage under the assumption it becomes the final repository.

13

Final Repository

This chapter describes the properties of the various barriers that prevent radioactive waste from reaching the biosphere. It emphasizes those circumstances that are important to the safety analysis with regard to the size of possible release. The purpose of contrasting reviewers' opinions is to demonstrate the extent of uncertainties regarding the properties of different barriers. Considering the demand for absolutely safety, it is, of course, the part of the uncertain interval in each barrier that might result in high radioactive doses that, if the correct value is there, is of major interest. The significance of such values, which cannot at present be precluded, is examined in chapter 14. This approach implies that those values, which the reviewers consider to be the "most likely" values, in many cases are different.

Glass

The safety analysis used the leaching time, that is, the total time from the initial start of leaching (when the capsule has been broken) until the glass block is completely dissolved. This gives the rate of release of radioactive materials into the groundwater. The leach time has been stated by KBS to be 30,000 years as a conservative estimate and 3 million years as a probable estimate.

Leaching times must be estimated on a basis of a combination of laboratory experiments and mathematical models, as field observations over extended periods of times naturally have not been made. No experiments have been made with cylinders of the size proposed by KBS containing the suggested quantity of activity (information from KBS to English and Lees). In laboratory experiments a specific leach rate is determined from a small glass piece. In order to calculate the leaching time of the canister an estimate of the exposed surface area is needed.

The leaching rate of the glass and the leaching of the radioactive materials depend on different parameters: the chemical composition of the glass, including its chemical variations due to radioactive decay; the chemcial composition of the groundwater and its flow; time; radiation; temperature; and pressure. No systematic experimental studies covering all these parameters have been done.

Leach rates reported in the literature vary over six orders of magnitude from 10^{-5} to 10^{-11} g/cm^2/d. In a Canadian field experiment a glass block was placed in earth and the leach rate was observed for fifteen years. The leach rate has dropped over time and the present rate is 10^{-11} g/cm^2/day. This was done with a glass type different from that proposed by KBS, and under conditions that are not representative for a repository. Therefore, it only suggests that leaching rates may be decreasing to very low values. Extrapolations of these results to other types of glasses and other conditions are not possible according to the National Physical Laboratory.

Several reviewers have commented on leach rates. KBS uses

2×10^{-7} g/cm^2/d. This is considered as probably realistic by Battelle NW, IAEA, and Choppin, for example, while PRAV and others believe it too high. Still others think this is an underestimate. National Physical Laboratory in England suggests 10^{-6} g/cm^2/d as do Rydberg and Winchester in their EKA review. Lawrence Livermore Laboratory and California Energy Commission suggest that 2×10^{-6} g/cm^2/d be used in the safety analysis. California Energy Commission says, "This contention is supported in KBS TR 50 itself, which states that the leaching results 'must be taken with considerable reservation' and 'continued studies on the behavior of actinide glasses will be needed to obtain a better understanding' so that better long-term predictions of their behavior can be made.' "

All these values refer to a temperature of 25°C during the leaching process. The temperature is a sensitive parameter. Leaching rates are expected to increase by a factor of 10 to 100 if temperatures increase from 25 to 100°C. The decay heat of the canister will increase temperature. According to KBS technical report 45, the temperature of the canister will be close to 40° after 500 to 1,000 years, diminishing to 25° after 3,000 years. That means, that for early breakdowns of the capsule and initiated leaching in the 1,000-year period an increase of the leaching factor due to increased temperature is motivated. The leaching rate at 70° has been measured over fifty days to be about 6×10^{-5} g/cm^2/d (KBS-TR-50) with a 9 percent waste glass. It seems that a factor of three would compensate for this increased temperature. All laboratory experiments have been carried out with much larger water flow than could be expected in a repository.

The leaching rates can be considered to be very tentative as experimental conditions are not representative of the actual situation to be encountered in the repository. For example, the glass to be used has not been available, so experiments have been performed with an inactive glass with the intended concentrations of nuclides in the waste. Furthermore, artificial groundwater has been used. The French glass that has been experimented on does not contain the amount of radioactivity intended in the KBS proposal.

The California Energy Commission comments, "Although solidified waste forms have been studied extensively, a committee at a Workshop on Ceramic and Glass Radioactive Waste Forms concluded that there is relatively little basic science data available to predict the long-term leaching of glass, glass-ceramic or ceramic waste forms."

In a memorandum Vattenfall agrees to the criticism of the leaching rates and considers a value of about 10^{-6} g/cm^2/d more likely to be conservative. The small amount of groundwater flow is considered, however, to be a limit and the lifetime of the canisters consequently millions of years.

The effective area available for leaching is another important parameter. The geometrical area of the cylinder is increased by a factor of five in the KBS report. This is done to include cracks in the glass. In the technical report discussing the GETOUT model, it is indicated that a factor of ten has been used.

Several mechanisms that could lead to increased surface areas are possible. Damaging could occur during handling, resulting in breaking the glass into pieces. In manufacture internal stress is likely to be built into the glass during cooling. Limited experimental evidence seems to be available to estimate the increase in surface area.

Battelle NW reports that a surface area increase by a factor of ten to fifteen has been observed using cylinders that are one third the diameter of the KBS cylinder. Other reviewers also suggest that a factor of ten to fifteen would be preferable. English and Lees from Jet Propulsion Laboratory consider the possibility of a very large surface area increase, up to a factor of 150 for the proposed canister size. This might be the result if the glass shatters due to tempering. This means that the glass can break into small pieces when stress is relieved. It is the same effect occurring, for instance, in car windows. Stresses are caused by temperature difference between the interior and the surface of the canister when the liquid glass cools into solid glass forms. No analysis of the stress distribution within the vitrified high-level waste has been made (statement from KBS reported in the review by English and Lees).

Lawrence Livermore Laboratory mentions that cracks in the glass cylinder might become filled with a hydrous gel that might reduce the area available for leaching. Battelle (SKI consultants) states that preliminary experiments have shown that micro cracks do not contribute to increased leaching, but that there need to be greater experimental efforts in order to be able to judge the effects.

Some reviewers point to the possibility mentioned by KBS that the glass block could be remelted in the steel canister to heal some of the cracks. It is, however, not known what that would mean to the canister integrity or to the internal stresses in the glass block. A Vattenfall memorandum states, without giving any references, that remelting prevents cracking. Another phenomenon that might contribute to cracking is the presence of seeds, that is, unmolten fragments of varying chemical compositions in the glass (statement from KBS related in the review by English and Lees).

The leach rate and the surface area available are inserted into a mathematical model to obtain the total leaching time. A few models have been proposed. In the first the lifetime is calculated in a leach-rate limited situation. In the second it is assumed that the leaching is limited by less solubility in the groundwater and that the water flow thus restricts leaching with a water flow rate of around $0.2 \, l/m^2/year$. This gives an estimated lifetime for the canister of 3 million years.

Rankama assumes that the KBS case with 3 million years is more realistic although he points to the need for better experimental data.

The National Physical Laboratory points to objections to such a calculation and suggests that it is unlikely that the flow will be as slow as that within the repository. Scientists there also point to the fact that no account is made of the fact that the rate of release of radioactive species is not limited by the solubility of silica alone, but more probably by its rate of hydration. Therefore it is likely that the real situation will be one of slow release due to hydration at the rate more characteristic of laboratory leach tests than one that in the long term will be controlled by silica solubility.

NPL obtains, with a leach-rate limited model with constant surface area, a total life of 10,000 years but reports it likely that a limited supply of groundwater will prolong the time to 30,000 to 50,000 years.

Other reviewers suggest shorter lifetimes. Rydberg/Winchester arrive at 6,000 years, English and Lees, 600 years, assuming the glass breaks into very many pieces enlarging the area by a factor of 150.

Luleå University points out that "*single* aspects of the glass and encapsulation integrity against corrosion and leaching are well studied but *the combined* corrosion effects from increased temperatures, radiation damages, and other parameters are less well studied. The stated encapsulation methods can, however, quite well prove to be practical."

It is also important to note that the experiments performed are very limited in time, only up to 100 days, sometimes up to a few years. All extrapolations of glass behavior over thousands of years are done from these limited data. These judgments are subject to considerable uncertainty due to limited knowledge. As correspondents for the American Physical Society conclude: "Our present knowledge of the properties of the borosilicate glass as a wasteform is inadequate to place reliance on the glass as the principal barrier to radionuclide release."

We therefore conclude that a considerable uncertainty exists due to limited scientific information and practical laboratory tests. If shattering due to the tempering process cannot be disregarded, it may mean that the leach time of the glass is as short as 600 years. Otherwise about 6,000 years leach time cannot be excluded.

Encapsulation and Corrosion

The glass is put into a capsule to prevent leaching of the radionuclides. This is especially essential during the first 500 years, as it is during this period that the dominating fission products, strontium and cesium, are decaying.

The capsule proposed by KBS consists of 3 mm of stainless steel, 10 cm of lead, and 6 mm of titanium.

In the general discussion on waste management and in some of the KBS review comments, nonmetals such as ceramics have been suggested for capsule materials. Such materials may be less susceptible to the chemical environment in the repository.

In KBS-II the lead-titanium capsule has been replaced by a copper capsule with an expected lifetime of 100,000 years. If this can be accepted after a technical review it will be of great significance to the safety analysis.

KBS states that the lifetime of the capsule will be at least 500 years but probably 1,000 years. Before a final judgment is done a more complete investigation should be undertaken. These judgments are, among other things, based on data regarding the size of the groundwater flow and its chemical composition which is obtained from other KBS expert groups. A Vattenfall memorandum claims that the lifetime could be further increased by adding iron(II)-phosphate to the buffer material.

As the review by the U.S. Geological Survey points out, the KBS report treats all the components of the waste containment system separately. However, they form a complex system that changes over time, and the behavior of such a system requires an exact knowledge of the waste form, including both the complex chemistry and valence state of the compounds. These data are apparently not yet available for the proposed scheme.

The National Corrosion Service at National Physical Laboratory comments that:

> it must be recognized that the proposals of the KBS report in the area of corrosion, and indeed we would assume in many other areas, go far beyond the currently available experimental data base. There may thus be many points that cannot be substantiated by recourse to fact of measurement and in such cases the criterion must be that a particular view is one that might reasonably be held after careful consideration by a majority of those well versed in the specific technical field in question. It is on this level of knowledge that many of our comments must necessarily be based.

Several types of corrosion have been investigated in the KBS technical reports. However, STU cautions that new types of corrosion do occur and it is hard to predict the corrosion in new environment. The University of Stockholm, for instance, points to the possibility of microbiological and bacterial corrosion.

Titanium

Several corrosion mechanisms for titanium are discussed by KBS and/or in the review procedure. They include uniform corrosion, local corrosion, stress corrosion, thermogalvanic corrosion, crevice corrosion, hydrogen-induced cracks, and others.

The experiments done to judge the titanium resistance by the KBS report are based on one-hundred-day experiments and twenty-five years of general experiments with the metal. Titanium has not been available in large quantities until recent decades. From this information the resistance of titanium is extrapolated over much longer time periods. There is some concern, expressed in the University of Lund review (Professor Östberg) and by the Royal Institute of Technology, that the protective passive coating of titanium oxide could fail. In such a case, a point attack could start after a long time.

Zirconium is a metal that is chemically similar to titanium. Zirconium has recently provided a surprise. After more than twenty years' experience a zirconium alloy has undergone local oxidation in nuclear reactors. The reason for this is not known.

Another type of local corrosion is the so-called sunburst. This is a local concentration of hydrogen that is taken up by the zirconium in the cladding to form a hydrate that is brittle and may initiate breaks.

Some uncertainties remain, according to FOA, regarding the resistance of the titanium capsule in the repository environment. It is principally possible that increased concentrations of fluorides and chloride ions in the groundwater might cause a somewhat increased tendency to corrosion. The National Corrosion Service also brings this up, stating that the lack of sufficient data prevents a quantitative evaluation. Furthermore, according to FOA, the uptake of atomic hydrogen might in unfavorable cases lead to the titanium cover becoming brittle and some cracks appearing. FOA points out that the overall effects of the taken standpoints might lead to a certain shortening of the capsule's lifetime in comparison with the assumptions in the report.

Stress corrosion, according to the KBS report, is theoretically possible but requires strong stresses in the capsule which can be avoided in manufacturing. However, according to KTH, that the titanium shell will be exposed to all the strain has not been taken into consideration. The lead will give no support, and inward bends in the titanium shell might easily result. This might lead to internal tensions of the same order of magnitude as the yield point of titanium. The National Corrosion Service also brings this up. It regards the data in this area as incomplete but thinks that the risk for further corrosion problems on that account is very small.

It is well known that titanium has a tendency to delayed fracture because of hydrogen uptake close to a fracture that is locally reducing the strength of the material. There are two theories based on limited experimental and theoretical work aimed at describing this phenomenon. It is, for instance, not well known whether a minimum hydrogen concentration is required to initiate this process. It is also uncertain whether hydrogen production may occur through reactions in or close to the titanium capsule.

In spite of these issues, the National Corrosion Service considers a capsule lifetime of one thousand years very conservative.

In its remiss answer to the KBS report the Corrosion Institute, the expert organ on which KBS based its recommendations, notes that delayed fracture following hydrogen uptake has a low probability but cannot be completely disregarded. Therefore, the titanium capsule cannot definitely be counted on to have a long lifetime even if it is likely to last for several thousand years.

Lead

The main reason for the 10 centimeter of lead in the capsule is that it forms a radiation shield, especially intended to reduce the formation of radicals in the surrounding groundwater.

Lead has a low melting point and is a relatively soft material subject to temperatures that can be regarded as relatively high. Lead could slowly flow under the action of mechanical stresses. The steel-clad glass cylinder might then float up to the top of the titanium canister according to KTH. This could be prevented if the lead is produced with large grain sizes, and it is therefore important that grain size after production is controlled. It is also important that lead not recrystallize later, which could happen if the canister is deformed. If the glass block floated up in the lead, the radiation protection of the lead would be reduced and chemically active species affecting corrosion rates might be formed outside the canister.

KBS argues that the transport time of lead away from the canister, once the titanium shield is broken, would be very long as this is controlled by the limited water flow and the solubility of lead. This would then give a very long lifetime to the capsule.

The California Energy Commission points to the possible formation of an intermetallic compound during lead casting or the welding of the titanium lid. Such a formation would weaken the outer can, so that it might no longer be self-supporting. If such were to occur, premature failure due to localized corrosion might occur, both in large numbers of canisters and in less than a thousand years. A Vattenfall memorandum considers such risk small as, when melting arcs, one has run across considerable difficulties achieving such compounds.

Conclusion

The Royal Institute of Technology concludes that a pessimistic judgment of the lead-clad titanium capsules' qualities suggests that its lifetime is uncertain and may be less than 500 years if one takes a pessimistic look on both the corrosion durability of titanium and the form durability of the lead cladding. As mentioned above, FOA also points to the possibility of a shorter lifetime than KBS has stated. In its remiss answer the Corrosion Institute gives lead in combination with the titanium a lifetime of at least one thousand years.

In summary, many reviewers think that the lifetime of the capsule will be thousands of years. There are, however, several points of uncertainty in the lifetime of the canister, and there are suggested mechanisms, which at today's level of knowledge cannot be ruled out, that might lead to a capsule lifetime even shorter than the minimum 500 years assumed in the KBS report.

Geology and Hydrology

General

A major barrier between the radioactive elements and the biosphere is the rock. When the capsule is broken and the water starts to leach the glass, the geologic barrier delays the entrance of the radioactive elements into the ecologic system. If a sufficiently long period of time elapses, then some of the radioactive elements will have decayed to harmless elements.

KBS has selected granite rock for the repository and pointed to three areas where it is believed a specific site (about one square kilometer) with acceptable properties could be located at a depth of 500 meters. In these areas one to three boreholes 500 meters deep have been drilled and some measurements performed in the holes and on the drill core.

The functioning of the rock barrier is dependent on the slow speed of groundwater motion and geochemical delays of the elements through absorption, precipitation, and other geochemical mechanisms. Also, the amount of groundwater flowing by the capsule is of importance for corrosion rates and the glass-leaching rate. It is convenient to look at groundwater movements and the geochemical retention phenomenon one at a time.

The discussion of the geology barrier has three parts:

1. What properties must the rock have to be regarded as an acceptable barrier?
2. Does a site with a rock volume of sufficient size exist with these properties?

3. Where can a place (geographically) be found with the properties in (1.) above, which by other reasons is acceptable as a repository for radioactive material (e.g., with no valuable ore)? How can a proposed site be shown to be acceptable?

The first point has not been discussed directly by KBS but is implicitly given as the values used in the KBS safety analysis. The third point is discussed in chapter 15 below.

In this section the properties that can be assigned to the rock barrier on the basis of today's knowledge and the experimental data presented by KBS will be discussed. This deals with the second point.

Four principle mechanisms may operate to bring the wastes to the biosphere from an underground repository:

1. Transport of the dissolved waste by groundwater
2. A geologic process that exposes the wastes, such as erosion, faulting, or meteoritic impact, which breaks the repository
3. A self-induced failure in which the mining or subsequent backfilling of the repository somehow fails and creates a short circuit to the biosphere, for example, failure of the sealing material in the shaft to the repository
4. Intrusion by man.

Item 4 above is not discussed by KBS and also not dealt with further here.

Winchester discusses the feasibility of geological disposal and gives a perspective on our current level of geologic knowledge:

Unlike the fields of physics, chemistry, and quantitative biology, fields which had already begun modern development by the early years of the 20th century or before, the revolution of the earth sciences came much later, largely since 1950. . . . Therefore, the earth sciences are "new. . . ." This leads to a level of certainty in the earth sciences which is much less than it is, for example, in theoretical physics.

Basic concepts in earth science have changed dramatically since 1950, and today we are witnessing an exciting period of earth science development that is comparable to the excitement we saw in the 1920s in physics and chemistry. On the one hand, this means that for the first time we are obtaining a much deeper understanding than ever before of basic processes in the earth, but on the other hand, as new concepts arrive each year old concepts of the previous year become severely modified or even abandoned. Therefore, our beliefs in the mechanisms of geology should be tempered by the realization that new evidence may arise in the near future showing them to be false.

The old concept of a static crust has been replaced by a dynamic concept including ongoing movements of several types. On the whole, earth sciences are in a dynamic stage of rapid change and progress.

The U.S. Environmental Protection Agency issued during spring 1978 a report entitled "State of Geological Knowledge Regarding Potential Transport of High-Level Radioactive Waste from Deep Continental Repositories,"[2] written by a specially assigned ad hoc panel of earth scientists. The panel was "surprised and dismayed" to discover how few relevant data were available on most of the candidate rock types including granitic types, basalts, and shales. The panel found the state of knowledge concerning total permeability in jointed shales, granites, and basalts still inadequate for meaningful forecasting. Total permeability includes all paths for fluid migration through the rock, including cracks, joints, faults, and inhomogenities of lithology.

Groundwater Flow

KBS has used two different approaches in discussing groundwater flow and the time for groundwater around the repository to reach the biosphere. One is based on a theoretical model using measured permeability values, and the other is based on age determinations of the groundwater from boreholes.

In its safety analysis KBS chose to use the "very conservative value of 400 years transport time in tight rock." The conservatism was primarily based on age determinations using the carbon-14 method. The theoretical calculations have been given less value.

The U.S. Geological Survey, in its review of the KBS report, maintains:

> Of the various mechanisms, transport by moving ground water is the most critical of failure modes for conditions discussed in the Swedish reports. All of the sites investigated indicate that the rock is fractured to depths of 500 to 600 meters. Measurements of in-situ permeability indicate that, while there are zones of intact rock with low permeability, there are also fractured intervals even at depth.
>
> The actual flow of water at repository depths will remain an uncertainty until more exploration of the sites is done and there is better understanding of flow in fractured media. The crystalline rocks considered have a wide range of physical properties. The fractures vary from barely perceptible to crushed zones one-half meter wide, some fractures have replacement minerals in them, such as smectite, carbonate, and quartz; others do not. A complete analysis of the perturbations to ground-water flow caused by the repository will require that a large volume of rock be characterized in detail.

The necessity of three-dimensional data on fractures and flow patterns has been stressed by many reviewers, among them, Scandpower, English and Lees, Dames & Moore, and the U.S. Geological Survey.

In a recent publication of the U.S. Geological Survey, several important criteria for adequate modeling of a high-level waste repository were stated:

> In order to predict contaminant movement, the physical and chemical properties of the media that control transport must be known for a length of flow path sufficient to describe the movement for the requisite radionuclide containment time. Although measuring these properties seems feasible, the manpower and observation time needed are not trivial. We need, as a minimum, the permeability and porosity of the media and the hydraulic head gradients all in three dimensions. In addition, we need to know the sorptive

characteristics of the media along all paths and we need to estimate the variable rates at which the solidified waste will enter the transporting fluids. Needed, in particular, is information on the distribution and extent of major heterogeneities. The need for such data severely taxes the available data base and the technology for generating it. Most of the requisite data are presently unavailable. Most of the available data have such large error limits that their usefulness in predictive models is limited.[3]

As demonstrated by English and Lees, the U.S. Geological Survey criteria are by no means fulfilled in the KBS report. (KBS replied to the question by English and Lees that they do not accept the USGS "minimum requirements.")

AIB/VIAK, consultants to SKI in their review of the KBS plan, state that it seems difficult to defend the choice of four hundred years transport time if the theoretical calculations are accepted.

These reviewers point to several of the KBS technical reports (KBS-TR-54:1-6) where transport times with brief durations on the order of *one year* are calculated for cross zones and on the order of *ten to one hundred years* for permeability values measured in the boreholes. However, the basis for calculations is weak. For example, data on the porosity of the rock are needed but in situ measurements of porosity are missing in the KBS investigation, as SGU points out.

CTH, like many of the foreign reviewers, thinks that the permeability of rock at depth of 500 meters should be further investigated. The method used by KBS, single borehole measurements, is certainly not sufficient to give even a general picture of flow paths in the rock, which was pointed out by many reviewers (e.g., Scandpower, English and Lees, Dames & Moore, Lüttig, USGS).

English and Lees, Dames & Moore, USGS, Scandpower, the University of Uppsala, KTH, and others criticize the KBS assumption of decreasing permeability with depth. They argue that this assumption is contradicted by the data presented in the KBS reports.

A permeability value of approximately 10^9 m/s was assumed for the granite rock mass at a depth of 500 meters. There is no substantiation given for assuming that the permeability does decrease with depth, and if the field measurements are regarded, this appears definitely not to be the case. The analysis assumes furthermore an anisotropic permeability (different properties in different directions). In this case the vertical permeability was assumed to be less than the horizontal. This is not supported by any data since boreholes were vertical and measurements of vertical permeability have thus not been made. Therefore this is not a conservative assumption.

A general problem is the treatment of fractures and fissures. These strongly affect groundwater flow, but their influence on permeability values is much smaller.

English and Lees give an example: the presence of a small crack might increase the average permeability by a factor of two, but groundwater flow is increased by a factor of 5,000. Thus, the permeability data give little information about groundwater flow.

The treatment of the permeability data is criticized by AIB (and by numerous other Swedish and international reviewers):

The permeability diagrams for the various boreholes often show sections of rock with very low permeability (lower than the measuring limit) but with a few higher values interspersed. The borehole in these sections might intersect a few widely scattered cracks with high water permeability. A system of such cracks might induce high water velocities and short transport times for the groundwater. They should therefore be subject to special analysis. It is inadequate to characterize a single crack with a k-value determined as an average for a $2-3$ m-long measuring section. It would be better to calculate the width of the cracks based on measured pressure and loss of water and use this value when estimating the water permeability of the rock and the transport time of the groundwater.

As an example, a 2 m-long measuring section with a permeability value of 10^{-7} m/s corresponds to a single crack of 0.06 mm width, assuming smooth crack walls. With a hydraulic gradient of 0.001, the water velocity in such a crack would be 3×10^{-6} m/s, which means that the water is transported about 100 m in one year.

The U.S. Geological Survey says in its review:

The approach that has been used to analyze the ground-water systems, utilizing porous media models, is appropriate at the sites investigated, given the little available data on hydrologic properties. However, the application of porous media models to simulate flow in fractured rocks requires careful interpretation of results. Results from porous media models can be used to calculate transit times for nuclide movement; however, these models provide averaged values that are not as meaningful as extreme values (which occur in large fractures with low resistance to fluid flow). The analysis by Neretnieks accounts for this variability, but his data on fracture widths are based on a very limited sample.

The absence of three-dimensional flow data in the KBS analysis has been criticized seriously by English and Lees, Dames & Moore, USGS, University of Minnesota, Scandpower, and others.

The California Energy Commission used Darcy's law and typical data presented in KBS-TR-30 in order to obtain an approximate transit time to compare with the 400-year transit time adopted by KBS:

Figure 6 in KBS-TR-54-06 shows estimated porosities varying between 1×10^{-5} and 6×10^{-5}; we adopted a value of 3×10^{-5}. We used permeability of 1×10^{-9} m/sec, the value that KBS used in the safety analysis. Section 5.1.2 of Volume II gives a hydraulic gradient of 8×10^{-3} meters/meter. Using these values in Darcy's equation given in Section 6.4.1 of Volume IV, we calculate a water velocity of 8.3 m/yr. If a single fracture were to connect the repository to a well or lake 1 km away, then the transit time would be 120 years. Slight changes in the values selected would reduce the transit time to less than 100 years. This simple analysis illustrates that the KBS choice of 400 years may not be particularly conservative.

The boreholes were taken to 500 meters depth, the same depth as used for the repository. AIB/VIAK observes that highly permeable rock may be found below a large zone of tight rock. Studies should therefore be extended to depths well below the repository. English and Lees, like the Geological Survey of Canada, question why boreholes were not taken deeper in order to rule out the possibility of regions of high permeability that are connected in several dimensions.

The considerable size of the area for the repository (1 x 1 km^2) in relation to frequency of fractures and bedrock composition was stressed by Dames & Moore.

According to USGS and Dames & Moore, it needs to be verified that the vertical permeability of the rock and the vertical flow paths induced by the vertical shafts will not result in shorter flow times than the 400 years assumed by KBS.

The U.S. Environmental Protection Agency's panel of geologists formulates two principal questions:

1. How does one determine real permeability of the rock mass surrounding the repository and extending to the surface?
2. How does the permeability of the fissures affect solute-retardation factors relative to the flux of water throughout the rock mass?[4]

The panel states that development of methods to answer the first question will be very hard, but must be undertaken.

The Norwegian Geological Survey, however, in its review, takes the position that the general geologic knowledge available is sufficient to give general predictions about the expected conditions of the actual depths and for the relevant times.

In conclusion, the way in which the presented permeability data have been used to determine the transport time of groundwater from the repository to the biosphere has been sharply criticized.

Groundwater Ages

The other line of argument concerning groundwater transport time is based on age determination of the groundwater by the carbon-14 method. At great depths, groundwater is expected to be quite old, but, as pointed out by the University of Gothenburg, KBS reports that groundwater ages of ten to twenty years have been measured in the Stripa mine at comparable depths.

Several reviewers have discussed the importance of the measured ages. It is noted that these measurements constitute the only data available in support of the hypothesis of very old groundwater and thus long transport times. It is noted, however, that what is measured is the time needed for the water to move to the sampling place, which may be the same but not necessarily the same as the time needed to move from the sampling place to the biosphere. The increasing temperatures with increasing depth act, for instance, toward decreased transport times in the upper flows.

The apparent ages of fifteen samples analyzed by the carbon-14 method fall in the range of 1,755 to 11,055 years. The apparent water ages by the tritium method for ten samples were 25 years or less. These later ages might, as stated by KBS, be due to surface water contamination of each sample by at least 25 percent. Depending on which component is measured—carbon-14 or tritium—different results are obtained.

To clarify this Winchester points out that:

> A simple model for resolving the desired "old" groundwater component could be a three-compartment model for mixing (a) young water, (b) old water, and (c) mineral constituents. The same model should be equally valid for the five components measured. Conceptually, the author of KBS 62 considered such a model although he did not discuss it explicitly or quantitatively. Nor did he show that a single model would be as a self-consistent one satisfying all five types of measurements."

A more realistic model would include the various ages of waters in the rock. Water moving in cracks is likely to be much younger than water in very narrow cracks. The measurements now give an average water age, but from the point of safety analysis, the faster-moving fraction would be of greatest concern and should therefore be calculated. Winchester states that until a model explains the relationships among various measurements quantitatively, the conclusion that an old groundwater component is present cannot be substantiated.

The data that have been presented and the interpretations made by KBS have been criticized. Professor Eriksson of Uppsala University states (in the EK-A material) that age estimates of groundwater from carbon isotope data in Swedish igneous rock areas are beset with some uncertainties that are not easy to resolve since the history of climate and land emerging from sea is involved also. The only safe procedure for estimating groundwater ages would be to sample in a borehole at different depths in an area with a vertical groundwater flow, and thus obtain values of all old water.

Professor Rankama from Helsinki and the Geological Survey of Canada both point to the need for more data to verify the KBS interpretations of the carbon-14 data.

SKI consultants, Scandpower, have phoned one of the world authorities on carbon-14 dating of water, Dr. W. G. Mook from the Netherlands. Dr. Mook reports that there are many unexplained and undefined questions with the method, and without further investigations he is not willing to give a definite statement.

In conclusion, the method of determining the transport time of groundwater to the biosphere based on carbon-14 data is regarded to have several uncertainties and its use by KBS is unsatisfactory.

Change of Rock Properties

The preceding discussion was based on the properties of the intact rock of today. Changes are conceivable, however, and it is necessary to evaluate the possibilities of such alterations as they would affect the safety analysis. Two main types of changes are discussed: those caused by geological factors and those caused by alterations of the rock during repository construction.

As stated in the AKA report, the stability of the Fennoscandian crust is the basis for a safe deposition within the bedrock. KBS calculated mean values of fracturing for the last 100 million to one billion years and used these as a measure of the crustal stability and a prediction of future geodynamic activity. This has been criticized by Scandpower, Dames & Moore, the Geological Survey of Canada, and also by Mörner in KBS-TR-18. Dames & Moore conclude that in the absence of data for a prediction of future deformations and stresses, "there is no basis for analysis of the potential for future disruption of the repository." Future seismicity is predicted on the basis of present-day instrumental records. This has been strongly criticized by Mörner (KBS-TR-18) and Tanner (Geological Survey of Canada).

The effect of ice ages on the repository has often been discussed. The last ice age ended about 10,000 years ago. The approximately 3 kilometer-thick ice depressed the rock by about 800 meters. The rock has since undergone an uplift of this magnitude and is continuing at rates to one centimeter per year. In two KBS reports different views are presented on the causes of the present uplift.

Mörner (KBS-TR-18) states that the Fennoscandian uplift is composed of two different factors: one typical glacial isostative that died out some 2,000 to 3,000 years ago and one linear uplift factor of uncertain origin that started at about 8,000 B.C. and is responsible for the present uplift.

Bjerhammar (KBS-TR-17) finds little support for this but believes that it would be extremely risky to use the crust for waste disposal if forces were operating to build up heavy stresses there. The University of Gothenburg would like to see a further review of the Mörner hypothesis, but otherwise the reviewers have not commented upon this problem.

It is generally believed that there will be new ice ages, and that, in fact, we may already be moving into the next one. Even if this is not the case, the time frame to the next ice age is in the order of 10,000 years. According to Mörner (KBS-TR-18), the peak rates of glacial isostasy (usually at the time of deglaciation and glacierization) are linked to intensive faulting, fracturing, and seismic activity. Despite this, KBS states that the dramatic change due to a future ice cover would not affect the repository. Dames & Moore discuss this matter and conclude that "the effects of future stress changes on the waste chambers in conjunction with a future ice age have been ignored." The U.S. Geological Survey states that "faulting and related fracturing associated with another ice age should be factored into the analysis." Conversely, Rankama thinks that the question of ice ages is academic. Possible damage to the repository occurs at a time with no population because of the ice cover. The length of an ice age cannot be predicted, and Scandinavia would be unpopulated for a long time.

There seems to be some uncertainty left regarding young faults and the probability of new faults. AIB/VIAK point to the limited investigations that have been made and the limitations of the methods used. For example, aerial surveys, the most commonly used method, are restricted in application by the presence of forests and interpretations are complicated. Also AIB/VIAK claim that there are about ten times as many faults per unit area as reported in KBS-TR-20.

Rankama states that there are neotectonic activities in certain parts of Sweden, but there are no indications of such activities in large parts of the country. It will be possible to locate a repository thereby avoiding areas with neotectonic activity.

Dames & Moore stress that the mere existence of the repository makes it "totally unreasonable to site the present fracture distribution in the bedrock of Sweden as evidence for future effects in the vicinity of a waste repository."

The KBS report states that new fractures and faults will occur at already existing joint planes, an application of the general rule that the weakest link in the chain breaks. To this the California Energy Commission replies that "If indeed the weakest link in the chain breaks during a seismic event, then the repository may be that weakest link."

KTH, Brekke, Fairhurst, Dames & Moore, and the California Energy Commission are concerned with the introduction of cracks in the rock walls of the tunnels due to excavation. Such cracks might open the possibility for rapid water movement along a tunnel, thus connecting to a shaft or high permeability area. Fairhurst argues that the fracture forecast should include mining and temperature effects, the latter due to the increased rock temperatures from the canisters. KTH says cracks of one to three meters will be formed during blasting. These cracks are small and therefore difficult to seal. Full-face drilling instead of blasting would decrease the crack formation considerably but a completely tight seal cannot be made, according to KTH.

English and Lees discuss temperature effects and refer to a May 1978 paper by Johnson and associates, where cracking of granite is reported to be observed during uniform heating of granite. The effect has been observed at 75°C for Westerly Granite. The reviewers report that this information was not known to KBS and that KBS should look into these possibilities.

USGS is skeptical about the piping system to irrigate the bentonite and suggests this could give rise to a "short circuit" and should be omitted if possible.

English and Lees are also concerned about the lack of depth in the discussion and the proposed plugging scheme for the tunnels and shafts. To what extent has bentonite been used before for such purposes and to what extent has its anticipated properties been verified? English and Lees put this question to KBS and comment in their review:

> Our discussion indicated that a bentonite-sand mixture had been packed in a large metal cylindrical container. Laboratory measurements of the physical properties of the mixture have been made including measurements of swelling. Neither small scale nor full scale experimental simulations including swelling of the bentonite caused by water injection have been made.
>
> The KBS staff stated that such simulations are not needed to demonstrate the adequacy of the proposed plugging scheme. Their reasoning is that the laboratory mea-

surement of the basic material properties and the "laws of nature" preclude failure of the approach.

This is an example of *technological optimism in the extreme*. Many, many projects have had disastrous failures even after much more careful analyses than that exhibited by KBS for the plugging. Here we conclude that the KBS position on the adequacy of the scientific data base for plugging is untenable.

In light of the experiments KBS has carried out up to the review, Dames & Moore comment regarding the plugging of tunnels and shafts with bentonite clay:

> Conclusive evidence of how a permeability (10^{-9} m/sec) of the backfill material will be achieved is not provided. . . .
>
> There is no demonstration that the backfill has adequate resistance to erosion by fluid flow. . . .
>
> The proposed technique of backfilling tunnels will result in a loose backfill. Evidence of sufficient swelling to ensure low permeability of the backfill is not presented.

Vattenfall in a memorandum does not share these conclusions and refers to the experiences in Sweden when using tightening earth material and to recent laboratory experiments.

Dilution Effects

The radionuclides leached from the waste glass and transported through the rock are assumed to be diluted in a recipient before being used as input data to the BIOPATH model, which calculates the resulting doses to man. The dilution effect is directly affecting the results of the calculations. It is therefore an important assumption, as scientists at the University of Uppsala have stressed also.

Several reviewers have commented on the dilution-in-the-well alternative. SGI regards the dilution model as very crude. Scandpower (consultants for SKI) points out that the radionuclides are assumed to be diluted in all the precipitation on a 2 km^2 large infiltration area. This is considered being optimistic and transport through cracks and other fissures may well lead to much higher concentrations, in some cases by a factor of ten, in a lesser volume of water.

The Geological Survey of Canada finds the dilution calculation "misleading" as it assumes dilution in the total yearly water input. In reality, the radionuclides would be discharged over a much smaller area: "the concentrations of radionuclides in these zones could be many orders of magnitude greater than those calculated based on dilution from water over the entire catchment area. Also, dilution in fractured rock would be very difficult to predict."

Conclusions

To summarize the geological discussions, the U.S. Geological Survey thinks the KBS plan is a promising concept that should be verified by continued experimentation, data collection, and underground testing.

A Vattenfall memorandum in its assumptions acts partly on the question of what properties a repository should have to be acceptable. This question is separated from the question of properties that can be shown to be achievable in a stated place. This regards the question *where* to locate a repository, which is discussed in chapter 15.

It cannot therefore be claimed that a place is identified which fulfills the properties that are assumed by KBS in its safety analysis. It cannot be precluded, according to several reviewers, that the transport times of the groundwater, in view of the presented data, can be assumed to be 10 to 100 years instead of the 400 years stated by KBS.

Retention of Radionuclides

Nuclides are delayed on their way through the rock owing to physical phenomena and chemical reactions. There are various chemical reactions, for instance, ion exchange, ion adsorption, reversible precipitation, and mineralization. These processes are collectively referred to as sorption and lead to a reduced transport velocity for radionuclides in groundwater.

This is described with a retention factor. If the retention factor is ten, this means that the radionuclide moves ten times slower than the water. For a water transport time of 400 years, a radionuclide with a retention factor of 100 would need 40,000 years to reach a recipient. According to KBS, it has not been possible to determine what processes are actually operating for the various nuclides. The processes are very complex and depend on the exact chemical composition of groundwater, rock minerals, and the introduced compounds from the waste glass and its encapsulation. Most reviewers stress that the laboratory experiments must be repeated and verified by experiments in natural environments in the field.

The U.S. Environmental Protection Agency Panel (EPA) summarizes current knowledge, stating, "The plethora of distribution coefficients and retardation coefficients data now available provide a good start, but they are not adequate to address the critical problem of retardation under conditions of high ionic strength."[5]

The sorption is often characterized by a distribution factor, called K_d (amount of radionuclide per kilogram of rock divided by amount of radionuclide per cubic meter of water). To estimate a retention factor, the distribution factor is determined in laboratory experiments. The retention factor K_i is then calculated

$$K_i = 1 + K_d(a_1/a_2)$$

where K_d is obtained experimentally, a_1 is the available sorption area in the rock (m^2 rock per m^3 water), and a_2 is the specific area of the rock sample used in the measurement.

As discussed in KBS-IV (page 66), there is some uncertainty regarding the areas a_1 and a_2. If the sample of rock in the experiment is considered as massive spheres, an a_2 value of 30 m^2/kg is obtained, but with a gas adsorption surface area determination method a value of 12,000 m^2/kg is obtained. Obviously, the same type of measurements must be done in the field as in the laboratory. However, this introduces an unquantified uncertainty. Another uncertainty is the amount of water present, which is needed for determination of a_1. This is coupled with the porosity that, as discussed above, has not been measured. However, Rydberg believes KBS has used a high porosity value leading to an underestimate of the retention.

Winchester has discussed the retention phenomenon in a statement to the Energy Commission:

> It should be emphasized that the retention factors are measured in the laboratory under idealized circumstances. In one type of experiment, pieces of rock may be put into contact with solutions containing various chemicals, including radioactive waste elements and the constituents of groundwater, and the transfer of radioactivity from the solution to the rock is measured. In another kind of experiment pieces of rock may be glued into tubes and a solution of the same constituents is pushed through the rock under high pressure. The solution emerging on the other side of the rock is analyzed and the extent of retention of radioactivity by the piece of rock is thus determined.
>
> Such experiments in the laboratory are difficult to perform and a very large number of different kinds of rock, water compositions, and radioactive element mixtures must be tested in order for definitive results to be obtained and the safety of a repository design to be verified. As I have reviewed the retention data produced by the KBS technical report series and additional literature published by investigators in other countries in the scientific journals, I have found that the current state of knowledge in general is very preliminary. In addition, I have found no direct experimental investigation of retention from solutions containing large amounts of lead such as discussed above. Consequently, the few laboratory measurements of mineral retention of radioactivity are woefully inadequate as proof of the

geological protection of man from the repository waste as in the KBS design.[6]

In KBS-TR-6, the state of the art of groundwater movements around a repository is discussed. The retention of radionuclides is treated, and it is concluded that

> since there are many factors that influence adsorption on rock surfaces and since saline water conditions cause complexities beyond those that are typical for simpler aqueous systems, there is no firm basis at present for developing predictions of radionuclide retardation in fractured-rock systems of the type that may occur at potential Swedish spent-fuel repository sites. Because values of K_a are not available for these conditions, inclusion of retardation terms in radionuclide transport models for the rock system cannot be expected to lead to very useful results. Unfortunately, there is no reliable basis for making even very general estimates of K_a values for fracture systems in granitic rock. Until this situation is rectified, all analysis of the consequences of migration of radionuclides from the earth materials into the rock system will contain large uncertainties.[7]

The California Energy Commission refers to the evaluation reported in KBS-TR-6, and concludes, "No information has been given that changes this picture, according to the Commission. Therefore, large uncertainties remain."

SGU considers the data for sorption relevant for the rock types and size fractions that have been studied. Field studies with natural groundwater are desirable to verify the laboratory measurements. SGU emphasizes the importance of the retention phenomenon and assumes its size will be further studied. AIB/VIAK say that KBS studies should be followed up with experiments where natural groundwater has been used instead of synthetic groundwater to check the effect of complex-forming substances on the sorption. Also, PRAV would like to see field experiments.

Rankama supports the conclusions drawn in KBS-TR-55 on which KBS has based its choice of retention factors.

Lawrence Livermore Laboratory finds the KBS values consistent with literature values and generally adequate for the predictions. Aikin and Hare also find agreement with reported literature values and believe underground experiments should be conducted.

Rydberg argues that the preliminary retention factors reported in KBS-TR-55 and used in the GETOUT calculations are too low since it has now been realized that the groundwater environment is diminishing, thereby lowering valence states of the elements concerned and increasing sorption. Also, sorption increases with time. More realistic values would be typically five or more times larger; for instance, Rydberg reports a neptunium k_d value increase from 40 to 2,000.

Choppin summarizes his review by stating that the measured, higher retention values reported by Rydberg would seem proper as they are reasonable and probably conservative measures of the retention likely in a repository. However, Choppin cautions that "major questions need to be answered about the effect of lead deposits of organic solubilizing agents of retardation release rates before the retardation can be assessed fully."

The studies referred to by KBS and Rydberg are based on two studies by Allard and associates. These authors (quoted in KBS-TR-98) caution

> The work was done under severe time constraints and therefore it was not possible to do many investigations. The following complementing studies would be of interest:
>
> 1. Studies of mechanisms of sorption
> 2. Analysis of chemical equilibria in groundwater
> 3. Studies of the dissolution of uranium oxide in water
> 4. Studies of organic-complex formers in groundwater
> 5. Investigation of the role of colloids for migration
> 6. Formulation of model and in situ measurements.

The U.S. Geological Survey states in its review that

More specific data on sorption properties of country rock are also needed. The various minerals lining the fractures need to be characterized and a thorough understanding obtained of the exchange reactions in which they may participate and of their distribution in the field. As noted above, the species undergoing transport may evolve with time and be quite different from those emplaced. Values used for sorption in the present model are probably accurate only to within 1 or 2 orders of magnitude.

The California Energy Commission adds that "The present capability for estimating the capacity of various geologic materials to retard nuclide movement is limited" (quoting U.S. Nuclear Regulatory Commission Staff Responses to Questions, 1977, pp. 1–143).

Sorption experiments are very sensitive to water chemistry and the concentrations of various substances in the water. In the laboratory setting the water chemistry and the concentrations of other materials can be carefully controlled. To be meaningful the laboratory experiments must duplicate the actual composition of the deep water found in the repository. It has already been noted that several reviewers severely criticized the carbon-14 method, in part on the basis that it could not be ruled out that the water samples partly consisted of surface water.

The size of the retention factors depends (KBS-TR-43) also on permeability and average distance between the fractures, that is, the transport time of groundwater. Shorter transport times for groundwater than those used by KBS in the safety analysis should thus mean lower retention factors.

Concluding Comments

The knowledge of retention phenomena is limited and mainly based on laboratory experiments. Recently published results from laboratory experiments (in connection with KBS-II, which has not yet been reviewed) suggest that the values used might imply an underestimate of the retention of the groundwater conditions assumed by KBS around the repository. These circumstances have, however, been questioned (see "Sensitivity Analysis of Some Coefficients in the GETOUT and BIOPATH Models" below). Several reviewers state that there are a large number of parameters that influence retention and that their mutual interplay is not fully known. This implies that the uncertainty about the retention values for a repository is large.

The BIOPATH Model

The BIOPATH model calculates radiation doses to individuals and also the collective doses from radioactive nuclides that escape from the repository and reach surface water. The GETOUT model calculates the quantities of radionuclides that reach the surface. The BIOPATH model uses the results of the GETOUT computation as input for the biological pathway calculation.

Three bodies of surface water are considered by KBS: a well in the vicinity of the final storage site, a nearby lake, and the Baltic. These are called the primary recipients. The radionuclides are then assumed to distribute themselves among seventeen various compartments, among them, water, sediment, soil, biota, and atmosphere. Thirteen different pathways, which experience shows are the most important ones for the radionuclides to reach man, are studied.

The intake of radionuclides by man is either through inhalation or through ingestion with food or water. From the concentrations of the elements in soil and water the uptake in vegetables and food, for instance, is calculated. There are about 1,000 coefficients in the model describing the pathways of the radionuclides among the different parts of the biosphere.

The BIOPATH model includes a number of simplifications: The system modeled is rigid, and the flow out of a reservoir is only dependent on the concentration of the nuclide considered; all reservoirs are momentarily well mixed; all atoms and molecules have the same probability of leaving the reservoir; and the transfer coefficients are time-independent. All these assumptions, of course, simplify the real conditions.

It is assumed that the radioactivity in the biosphere is diluted with assumed amounts of water. This gives a certain concentration of radioactivity in the well and in the lake. Depending on the assumptions made of the hydrology in a region, the activity may be diluted with other quantities of water than those used in the computations. As discussed in the geology section, several reviewers are skeptical of the anticipated dilution as being a conservative assumption. According to Scandpower's report to SKI, a dilution of the activity in the well case could be such that only one/tenth of the used value could occur, resulting in a tenfold larger exposure for a critical group in the well alternative.

Two principal questions must be asked regarding the BIO-PATH model:

1. To what extent does its structure give a realistic description of the processes of interest for the dose calculation?
2. Are the coefficients and other assumptions well founded?

Regarding the first question, IAEA is satisfied that the model used in the assessments is adequate and concludes that the assessments incorporate a substantial margin of safety. Also, University of Uppsala finds it likely that more realistic assumptions would give considerably lower dose contributions than those obtained in the model.

Aikin and Hare find that BIOPATH is ". . . a crude model, given the complexity of ecosystems functions. In particular, the reservoirs are oversimplified."

However, these reviewers find the overall result of the KBS project to be well on the safe side.

The California Energy Commission, by way of contrast, writes in its review:

KBS maintains that the transfer of radionuclides from one compartment to another and uptake via food chains can be calculated through the use of transfer coefficients. This may be true in principle but is certainly not true in practice. A perspective on the problem, different from that given by KBS, is given in a U.S. Environmental Protection Agency (EPA) report:

'Variability in the concentration factors reported for similar organisms from different but similar environments appears to preclude the use of general concentration factors to assess the current or previous levels of radionuclides in any particular aquatic environment. Although many freshwater and marine organisms accumulate or concentrate radionuclides present in the aquatic environment, the concentration factors reported in the literature indicate that *local conditions strongly affect concentration factors for organisms.* Even for a given area the concentration factor for a given organism can be expected to vary with changing local conditions; for example, season, water temperature, and total biomass present' (Ref: Robert G. Potzer, "Concentration Factors and Transport Models for Radionuclides in Aquatic Environments." U.S. EPA Report EPA-600/3-76-054, May 1976, p. 3, emphasis added).

Considering these variabilities and the fact that KBS used only 13 pathways, it is not clear that they adequately treated the significant cases.

Studsvik Energiteknik AB states that these uncertainties of the concentration factors have to a certain degree been considered by often choosing high factors.[8]

Dr. George Woodwell, Marine Biological Laboratories, Woods Hole, Massachusetts, while finding the time allowed for a detailed review of the KBS work too short, states in general that:

The model provides at best an estimate of how the world is likely to work. Proof that the model is correct and yields a prediction of high value hinges on repeated use in circumstances where the accuracy of the model can be checked. In the present instance there is no way of testing the accuracy of the model in predicting the movement of radionuclides over thousands of years. The best we can do is to examine all of the assumptions, test all of the inputs and examine the sensitivity of the conclusions to the range of assumptions and numerical values that seem reasonable at present in approaching the problem of predicting hazards to man. While we think we know the major pathways for short-term accumulation of most radionuclides in the biotic systems, I am certain that we will discover, if we can stay at it long enough, that there are many other additional pathways for transport of long-lived radionuclides when we consider periods of centuries to milliennia or longer. This observation simply suggests that any prediction is bound to be tenuous if the prediction spans a period of a thousand years or more. . . .

There is neither agreement nor a generally accepted protocol for describing the movement of radionuclides through the environment. The technique used in BIO-PATH has relied heavily on accumulation factors. While this approach is a reasonable first step, it is far from satisfactory for long term prediction.

To the California Energy Commission the volume of water or soil available for uniform mixing and dilution appears to be unrealistically large for the intermediary, regional, and global ecosystems. For example, the model includes the assumption that radionuclides will mix uniformly throughout the Baltic Sea. Similarly, the population density is taken to the average Swedish value.

Whether or not the model and coefficients are well founded, Bertil Persson and Mats Nilsson, reviewing the BIO-PATH model for SSI, write, "It would have been desirable to have access to some documentation of the computer code used in the model. In this KBS/AE could for various reasons not oblige."

By performing independent calculations with an alternative mathematical method, however, the two reviewers verify the mathematical solutions reported in KBS-TR-40. Nevertheless, when large differences between input and output rates for a compartment occur, the KBS mathematical solution is less exact. This may give a deviation of a factor of two or less in some compartments, but it will not change the overall results significantly.

University of Lund points out that when choosing values for uptake factors and concentration factors in KBS-TR-40, the lower values have been preferred in some cases without any justification, for instance,

	Cs-135 and Cs-137 in literature data	Used type values
Uptake by plants-soils	$3.10^{-3} - 7$	3.10^{-3}
fish-lake	$500 - 12{,}000$	$2{,}000$

The choice of uptake factors is particularly critical for Cs-135, as this radionuclide gives among the highest individual doses. With a tenfold increase of the uptake factor for Cs-135 the same values in the highest individual dose, as in the well alternative, would be reached in the lake case. The collective dose will also be somewhat higher as more individuals are affected in the lake case.

Still, SSI emphasizes that KBS, faced with a choice among various parameter values, often has used those that give the highest radiation doses in their calculations, for example, sedimentation rates of the radionuclides in the oceans and for some habits of the population that would expose individuals to external radiation, such as when handling contaminated fishing equipment. However, such external exposures will not contribute significantly to the collective doses. Similarly, SSI discusses the influence of changes in food consumption. It cannot be ruled out that man will be exposed through new means, such as intake of desalinated ocean water or ocean products such as krill, octopus, and algae. Such changes, which are not unlikely —that 20 percent of meat consumption is replaced with algae (10 kg/ind year)— would increase the collective doses.[9]

The calculation of collective dose commitment as suggested by KBS indicates a value of 300 manrem/MW(e)a integrated to infinity (according to SSI's remiss answer). SSI has commissioned alternative calculations of the collective dose commitment. Using a model originally developed for the Argentina Atomic Energy Commission, values as low as 1/10,000 of the KBS values were obtained. Even with changes in diet patterns, the highest values were only 1/200 of those reported by KBS. The main reason for this discrepancy, according to SSI consultants Persson and Nilsson, is the treatment of transfers to and from the ocean sediments. In the BIOPATH model, radionuclides are assumed to return to the water from the sediment at a certain rate. This phenomenon has been observed in field experiments. In the Argentine model, this process has not been incorporated, which leads to lower radionuclide concentrations in the water than obtained with the BIOPATH model.

In general, little is known concerning the behavior of released transuranic elements into the sea. This is particularly true for neptunium, the element that would be responsible for the highest global dose commitment. Reported literature values of biological uptake of transuranic elements differ by several orders of magnitude. KBS tried to overcome this situation with an overconservative sedimentation model, according to Persson and Nilsson. The effects on the well and lake cases are less clear as to the effect of these uncertainties.

The results of the estimates in this report differ in certain parts from KBS estimates as changes have been made in the BIOPATH model. The changes that have been made (by Studsvik Energiteknik AB) in order to increase its relevance imply, among other things, that the radiation doses received by individuals in the lake alternative have increased by a factor of thirty-three for Pu-239.

In summary, the model itself has been little criticized except for documentation of the computer code. Some doubt exists though as to the applicability of the model for large compartments and for very long-term perspectives. The largest uncertainty is the supply of data especially with regard to concentration factors. This suggests that caution in interpreting the results is well justified.

Radiation Exposures from the Repository According to KBS

The KBS report calculates radiation doses for three cases: radioactivity leaking to a well, to a lake, or to the Baltic. Results are given for the largest individual dose obtained by the most-exposed individuals (the so-called "critical group"), and the collective dose. The individual dose is of interest to estimate the risk for the most-exposed persons of receiving, say, radiation-induced cancer. The collective dose is of interest to estimate the number of cancer cases and genetic damages in a large population over long periods. The collective dose is the sum of all doses received by all individuals exposed or, what amounts to the same thing, the number of individuals times the average individual dose for the individuals under consideration. The unit for collective dose is thus man times rem, manrem. It is normally reported in relation to the amount of electricity generated expressed in MW and years, MW(e)a, or megawatts of electricity per annum. "All individuals" are then taken as the world population (assumed by KBS to be 10 billion people eventually).

The main case reported by KBS is based on a capsule lifetime of 1,000 years, a glass leaching time of 30,000 years, a groundwater transport time of 400 years, and retention factors according to table 6.1 in KBS-I-4.

The individual doses are for the various cases:

Case	Dose rate	Converted to dose/year
Well	0.4 rem/30 years	13 mrem/year
Lake	0.02 rem/30 years	0.7 mrem/year
Baltic	0.0002 rem/30 years	0.007 mrem/year

The differences among the cases are large, with the Baltic 1/2,000 of the well case.

The natural radiation background in Sweden is about 3 rem/30 years (100 mrem/a).

The collective doses are reported for the 500 years of highest exposure. The results are:

Case	500-year collective dose
Well and lake	0.007 manrem/MW(e)a
Baltic	0.006 manrem/MW(e)a

The recommended limit for collective doses is 1 manrem/ MW(e)a for the total nuclear fuel cycle. About half of that is allocated to power station operation, and a similar amount is expected for reprocessing. The values reported by KBS for the final repository would therefore not be the most critical in the fuel cycle.

Figure 13.1 shows the contributions of the different nuclides to the individual dose in the well case. As can be seen, neptunium contributes most, followed by technetium, radium, uranium, and cesium. In the other cases, neptunium and cesium dominate. The largest contributors to the collective doses are iodine, cesium, neptunium, thorium, and technetium.

KBS alleges the main case (summarized above) is very conservative. To illustrate the degree of conservatism, a case is calculated with a glass-leaching time of 3 million years (it is argued that the amount of water present is so small that glass solubility will limit the glass leaching) and a groundwater transport time of 10,000 years (it is argued this is an indicated value from the water age determinations).

The retention factors for neptunium and plutonium are increased furthermore by a factor of ten (it is argued recent laboratory experiments indicate this is likely). With these assumptions, the individual doses in the well alternative are reduced from 0.4 to 0.001 rem/30 years.

KBS states a probable glass-leaching time of 3 million years

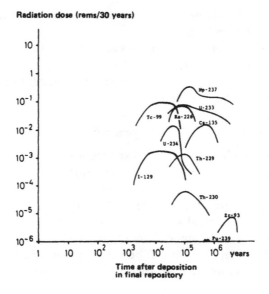

Fig. 13.1 Maximum doses from various nuclides (calculated for individuals in critical group). Calculations apply for a well as recipient and a glass-leaching period of 30,000 years (source: KBS-I, IV).

and otherwise the same conditions as in the main case. This gives a dose of 0.007 rem/30 years in the well alternative.

KBS summarizes the situation in figure 13.2.

SSI states that several reference values as given in the KBS report (for instance, in figure 13.2 above) are misleading or incorrect:

ICRP's limit for individuals exposed during a succession of years is, not 500 mrem/year, but 100 mrem/year.

The highest value for radium in drinking water "from a single rock well," which is stated to be 400 mrem/year, is related to water in a mine and not to common drinking water even if it is used as such by the mine staff.

The value stated for "radium in drinking water" is an order of magnitude too high. In the table on page 119 in Part I, the

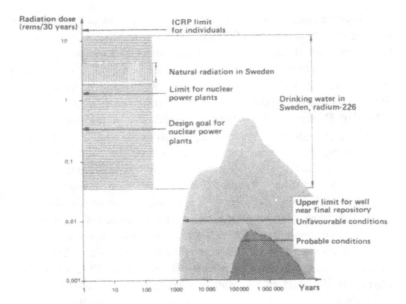

Fig. 13.2 Radiation doses (upper limits) to critical group. Calculations pertain to slow decomposition of canister with a well as primary recipient. For comparison, dose load from natural radiation sources as well as established dose limits have also been plotted (source: KBS-I, IV).

natural radium content in Swedish drinking water is stated to be 0.1-40 pCi/1. The highest value concerns the above-stated mine water. The highest value SSI has measured in any municipal drinking water is 10 pCi/1 but the average lies below 0.3 pCi/1.

The collective doses discussed above are calculated for a 500-year period. SSI objects to this method of using the collective-dose concept but finds it giving a correct number by coincidence.

During its review, SSI asked for calculations of collective doses summed over longer periods, including a summation to infinity, that is, the time when all nuclides have disintegrated through radioactive decay. KBS reported for this case a value of 300 manrem/MW(e)a. After 100,000 years, only 0.15 percent of this dose has been received, that is 99.85 percent of the damages will come only after 100,000 years. It is argued by SSI though that the model used for this is inadequate, as it does not properly handle the role of ocean sediments. The removal of nuclides to the sediments is underestimated, and with another assumption SSI arrives at collective doses to infinity around 1 manrem/MW(e)a. However, the group in Studsvik, which developed the BIOPATH model, does not agree their model is inadequate considering the present state of knowledge.[10]

In calculating the collective doses, it has been assumed that carbon-14 and most of iodine-129 are removed in the reprocessing step with only 1 percent of the iodine remaining in the glass. The contribution from all iodine, were it all to be released, would therefore increase by a factor of 100. The 500-year collective dose would then increase to 0.6 manrem/MW(e)a. SSI believes, for instance, that the very long half-life of iodine-129, 17 million years, would lead to its release sooner or later. However, the individual doses from iodine would reach 0.2 rem/30 years during the period 20,000 to 30,000 years after disposal in the well case. In this more realistic Baltic case, the value would be 5×10^{-5} rem/30 years. FOA notes that the I-129 values reported by KBS appear large compared with values reported by UNSCEAR. The 500-year collective dose from carbon-14, according to SSI, would be 2 manrem/MW(e)a.

SSI thinks it is desirable in the long run to remove the carbon-14, but considering its half-life of 5,700 years and its chemical properties, SSI does not consider that keeping it out of the biosphere should be a large problem.

PART IV

Sensitivity Analysis and "Where" the Wastes Can Be Finally Stored

14

Sensitivity Analysis

The previous chapter concluded with a statement concerning the safety level of KBS final deposition, according to KBS's own analysis. The discussion derived from results achieved using GETOUT and BIOPATH, the mathematical models explained in chapter 13. Many parameters are included in these models. The most important are the time for capsule disruption, the leaching rate of radioactive materials from the waste glass, the groundwater transport time from the repository to the biosphere, and the retention factors for the various radionuclides.

All these parameters are not well understood. There is thus an uncertain interval within which the "correct" value for a certain site is located. In order not to underestimate the possible effects of the final repository on man, these parameters therefore must be given values that allow for the uncertain intervals. In chapter 13 we stated those points of view that were aired during the review process; the possibility that the KBS parameter values do not fully cover the uncertain intervals was discussed.

The issue to which sensitivity analysis might contribute is whether or not such uncertainties are critical in determining the safety level of the final repository. The radiation doses to which man would be exposed based on changed parameter values are calculated in this chapter. It remains then to decide whether the calculated safety level can be accepted. This implies a judgment of whether the parameter values used by KBS can be set aside and further, whether the radiation doses are socially and scientifically acceptable.

The models that were used by KBS are the ones used here. The computations were done by the same persons who did the computations for the KBS project.

Description of Analysis

Sensitivity analysis is a purely mathematical exercise. In the mathematical models used to calculate doses from the repository to man, a large number of simplified assumptions have been made which are necessary to make a model of reality less complicated than the reality itself. Sensitivity analysis gives no information on how well a model really represents reality, that is, its relevance. Sensitivity analysis does tell us, however, about the importance of various numerical assumptions made in the models or in the input values. Necessarily sensitivity analysis takes up a number of combinations of parameters and aspects.

The calculated main cases are as follows:

The KBS Cases: Assume the Same Parameter Values as KBS

The parameter values stated by KBS in its main case are here referred to as "KBS main case." Disadvantageous conditions were assumed according to KBS. Another case, "KBS probable," also was presented.

A Cases: Assume a Single Barrier Fails

One of the barriers (for instance, the leaching time of the glass, the capsule, or the geology) will perform badly. In the analysis it is given a parameter value that, with today's knowledge, cannot be totally excluded but which does not seem very likely. The other barriers are given favorable values in this case.

B Cases: A Conservative Case

The values chosen for the parameters in this case are, according to the reviewers, within the existing intervals of uncertainty. The barriers are not independent of each other. For instance, the groundwater flow is of importance for leaching of glass, corrosion of the encapsulation material, and the transport time for groundwater from the repository to the recipient. Two subcases, B 1 and B 2, have been calculated. The first merely assumes use of uranium fuel, the second assumes use of plutonium in the fuel (see chapter 5).[1]

C Cases: Assume a Geological Event That Strongly Affects the Final Disposition

Assume that the final deposition is intact for 10,000 or 100,000 years followed by a geological event that causes short transport paths to the biosphere, equivalent to case B 1.

D Cases: Use of Plutonium

Assume that the plutonium from reprocessing is used in light-water reactors.[2] Otherwise use the same values used by KBS in its safety analysis.

The values used for the various parameters or barriers in the different cases can be seen in table 14.1.

The calculations stated below (in the section entitled "Estimated Radiation Doses") were made with the same coefficients and parameters used by KBS. Since KBS made its calculations,

TABLE 14.1
CASES USED IN THE SENSITIVITY ANALYSIS
(in hundreds of years)

	Notation	Capsule disruption/year[1]	Glass-leaching time/year	Groundwater transport time/year	Retention factors (F-values in KBS/IV)[2]
A) One of the barriers performs badly:					
A1) Short groundwater transport time, long glass-leaching time	A1	5	3,000	0.10	F/10
A2) Short glass-leaching time, long capsule disruption and groundwater transport times	A2:1	50	6	100	F
	A2:2	50	20	100	F
A3) Early capsule disruption	A3	1	3,000	4	F
B) Coupled parameters, considering uncertain knowledge					
B1) Only uranium in the nuclear fuel	B1	5	60	0.40	F
B2) Reprocessed plutonium and uranium in the nuclear fuel	B2	5	60	0.40	F

C) The repository intact for some time, thereafter high groundwater flows caused by some geologic event				
C1) The repository intact for 10,000 years	C1	105	60	0.40	F
C2) The repository intact for 100,000 years	C2	1,005	60	0.40	F
D) Use of reprocessed plutonium and uranium in the nuclear fuel				
D1) Parameters according to KBS main case	D1	10	300	4	F
D2) Parameters according to KBS probable	D2	10	30,000	4	F
KBS values according to KBS both main cases				
KBS main case	KBS main case	10	300	4	F
KBS probable	KBS prob	10	30,000	4	F

1. The stated values should be compared with those discussed in part III.
2. Different retention factors for different nuclides.

certain modifications of the models were made by KEMAKTA
and by Studsvik Energiteknik AB, respectively. In order to draw
a complete parallel, new calculations using KBS parameter values
have been included also in "Estimated Radiation Doses."

The analysis in this chapter is not complete. Not all parameters have been tested, and no systematic investigation of the
coupling effects between changes in various parameters has been
performed. Such an analysis would have been desirable but was
not possible because of time constraints on the present work. The
sensitivity analysis used the same method used by KBS, aimed at
choosing disadvantageous values for the various parameters in
order to achieve a conservative estimate. Another approach is to
simulate input on a statistical basis. The probability of various
parameter values can then be used. In such an analysis the
parameters and coefficients in the model vary, and the value is
assigned on a probability basis. By performing many (thousands
of) calculations a probability distribution of the resulting radiation doses can be generated, shedding additional light on the
importance of the uncertainties in making judgments.

Estimated Radiation Doses

Maximum Individual Doses to Critical Groups

Figure 14.1 gives the maximum exposure to individuals in the
various cases if, in the future, a well is drilled in the vicinity of the
repository. Figure 14.2 shows the individual doses in the case
where the radioactivity is leached to a lake in the repository area.*

*Maximum individual dose is expressed in the figure as the value for the nuclide
that gives the largest dose rate. The time available has not allowed the dose rates
from the various nuclides to be summed over time and the total individual dose to
be stated. This implies that the stated values are an underestimate, but the
divergence is limited for most cases. In case B1 (figure 14.1), for instance, 100
mrem/y is the dose rate stated for Pu-239, while the total dose rate from all
nuclides is 110 mrem/y. The reference values from ICRP and for drinking water
differ from those stated by KBS; they are here illustrated according to SSI's remiss
answer.

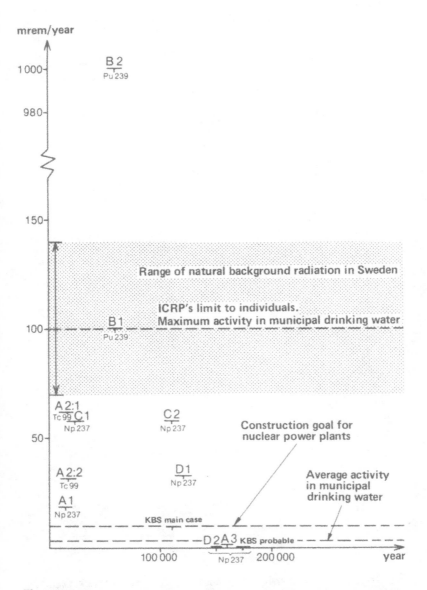

Fig. 14.1 Maximum annual individual doses to critical group in the well alternative.

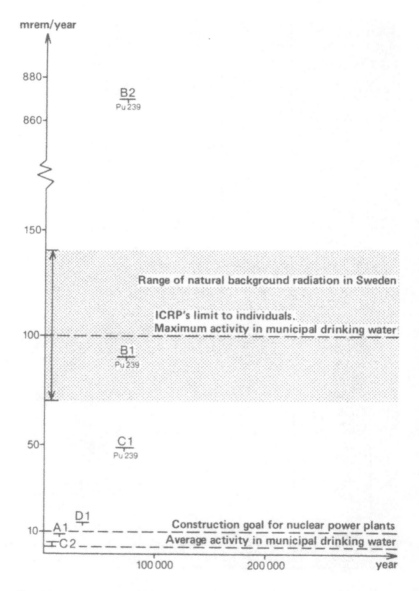

Fig. 14.2 Maximum annual individual doses to critical group in the lake alternative. Estimated cases giving less than 3 mrem/y are not marked in the figure.

The theoretical radiation doses are distributed over a period determined by the rate at which the radioactive materials reach the recipient and the nature of their movement through the biosphere. This time distribution is illustrated in figure 13.1. For each calculated case only the peak value for the radiation dose and the time it is reached are stated in the figure.

The A cases illustrate what happens if it turns out that one of the conservative values picked by KBS is optimistic. The other parameters for this exercise have been given relatively propitious values.

The figures show that the radiation doses vary among the calculated cases. Some of them are in the same range as KBS main case, some are below, and some are above.

The B cases illustrate a combination of values that from the review procedure appears possible but not necessarily likely. The result is 100 mrem/y, doses eight times higher than reported by KBS (main case) as a conservative upper limit when the nuclear fuel consisted only of uranium. It is the same level as ICRP's limit value for individuals who during a succession of years receive high radiation doses. If plutonium is used as nuclear fuel, the values will be ten times higher, that is, the individual doses will be about ten times higher than ICRP's limit value.

The C case anticipates an unlikely but not totally dismissable geologic event that occurs in tens of thousands of years and implies that the repository is exposed to an increased ground-water flow. Doses (60 mrem/y) will then exceed those predicted in the KBS main case and will be about half of ICRP's limit. A meteoritic impact qualifies as such an event.

Case D is the case where the plutonium extracted during reprocessing is used as nuclear fuel. This increases the amount of some actinides (transuranic substances). The final repository is supposed to have properties according to "KBS main case" and "KBS probable" respectively. The results of the use of plutonium are a threefold increase of the doses.

This last variation, a factor of three, might in some cases also be superimposed on cases A and C (still not A2:1, A2:2, where the doses are dominated by a fission product). This uncertainty is solely due to a decision to recycle the plutonium. If reprocessing takes place, the plutonium must be stored, and as a result an estimate of the release from storage must be made. This is not described in KBS-I.

Another factor that affects all of the cases is the choice of dilution in the well, the lake, or the Baltic. As discussed in the geologic section ("Geology and Hydrology") several reviewers point to the possibility that dilution would be decreased by a factor of ten, which would increase doses by a factor of ten for individual doses in the well case.

Another variation that might be considered results from uncertainties in the BIOPATH model. As discussed below ("Sensitivity Analysis in Some Coefficients in the GETOUT and BIOPATH Models") this might introduce an additional factor of two or three.

If these parameters (dilution and coefficients in the BIO-PATH model) were changed in order to consider the uncertainties in the data, this could mean in disadvantageous cases that the radiation dose in cases A, B, and C, the well alternative, would be more than ten times higher than stated in figure 14.1.

Consequences of a Nuclear Era on the Future Release of Radioactivity from Repositories

In order to achieve an upper limit on the releases from nuclear power, it is assumed by the regulatory agencies that the nuclear era will last for 500 years and that electricity consumption will be 10 kW per capita (which is more than five times higher than today's level in Sweden). Using these assumptions, an acceptable release from a single power station can be calculated. In the case of operation of nuclear power stations, the yearly released activity is proportional to the total installed capacity. In the case of future releases from nuclear waste repositories, the releases are proportional to the total accumulated quantity of radioactive material produced during the nuclear era. This amount is proportional to the total capacity multiplied by the operation time.

In order to illustrate the consequences of a large global nuclear program, we can assume for a period of 500 years a nuclear capacity corresponding to 10 kW(e) per capita. With this information the average exposure at a future time (10,000 years) to every individual can be calculated. This is a way of describing the limited effects of the nuclear era for an average radiation dose. An analogous argument was used by SSI in its remiss answer, but they have used instead the concept "limited collective dose commitment over 500 years." This collective dose commitment shall be below 1 manrem/MW year to give less than 10 mrem/year average dose commitment to the world population, which SSI considers to be acceptable. The limit of 10 mrem/year can be compared with the values given in table 14.2.

It should be noted, however, that all this exposure cannot be allocated to a final repository of highly active waste. Other long--lived radioactive wastes that might be released during other portions of the nuclear fuel cycle (for instance, uranium mining and reprocessing) would also contribute.

These values expressed in table 14.2 refer to average exposures, when the released doses are spread over the entire world population. Critical groups (e.g., those eating fish if fish is a dominant source of exposure) will receive doses that are much higher. Such heightened exposure may be 100 to 1,000 times as large as the average exposure in table 14.2. The relation between the exposures to the critical group and the average population depends on many factors; it is hard to make a precise estimate. A comparison of the individual doses (given a certain amount of released radioactivity) with two of the cases calculated with the BIOPATH model gives a difference between the well case and the Baltic case of a factor 700 to 8,000 (depending on nuclide).

Table 14.2 shows that the average dose to the world population in KBS main case is 0.4 mrem/y, clearly below 10 mrem/y. If the uncertainties of the constants of the models are considered, the value is 4 mrem/y. If the values of the various barriers that cover the uncertain intervals given by the review (case B) are considered, we obtain when using uranium fuel (case B 1) 3.7 mrem/y, and 4 mrem/y if the constants in the model are changed. When recovering plutonium and uranium (case B 2) we obtain 35 and 39 mrem/y, respectively. At the same time the individual doses to the critical group are considerably higher.

TABLE 14.2
AVERAGE INDIVIDUAL DOSES TO THE MOST
EXPOSED FUTURE GENERATION
(based on a large global nuclear program)[1]

Releases from waste repositories that all have qualities according to case	mrem/year global average to individuals	
	Original model[2]	Changed coefficients[3]
KBS main	0.4	4
KBS probable	0.016	0.16
A1	0.7	0.8
A2:2	0.36	—
B1	3.7	4
B2	35	39
C1	2	2.2
C2	0.34	3.4
D1	1.6	16
D2	0.054	0.54

1. The calculations are as follows: Assuming a nuclear era of 500 years and an installed capacity of 10 kW per capita which gives a total produced energy of 5 × B MW–years, where B is the world population which is assumed to be constant. If the yearly total dose commitment from a waste repository is X manrem per year and MW–year, then the average dose to an individual can be calculated as $5 \times B \times X/B = 5 \times X$ rem/year.
2. Calculated with the GETOUT and BIOPATH models.
3. Changes in the concentration and uptake coefficients which are discussed in "The BIOPATH Model" below.

This illustrates a situation where a repository of KBS type was installed on a large scale over a very long period. No consideration had then been given to improvements that can be made during the period regarding construction and knowledge, and thus this only gives a hint of the extent to which present knowledge can make a sufficient basis for judgment today as to whether a 500-year program could be accepted.

Total Collective Dose Commitment

The discussion of acceptable, future radiation exposure from final storage of radioactive wastes has just started (see chapter 8).

Figure 14.3 shows the total dose commitment per MW-year, that is, radiation doses summed over all individuals and over time. The calculations are made using the BIOPATH model. It has to be recognized that the pathways of radioactive elements in the biosphere over hundreds of thousands of years are not well understood. The numbers are therefore uncertain and some, including SSI, believe that they are much too high (as discussed in chapter 13 under the heading "Radiation Exposures from the Repository According to KBS").

There is a standard for radioactive releases from the total fuel cycle: the total dose commitment shall be below 1 man-rem/MW(e)year. As can be seen in figure 14.3 most of the calculated cases exceed this figure when the radiation doses are summed over periods that exceed tens of thousands of years. This standard was developed considering short-lived radionuclides from nuclear power plants. Whether it should also be applied to release of the long-lived radionuclides, produced as a consequence of the use of nuclear power, must now be discussed.

Summing the collective doses over the next 10,000 years shows that the total dose for the KBS main case is 0.06 manrem/MW year (0.23 according to SSI's remiss answer). If the sensitivity analysis case B is used, a dose commitment of 0.27 manrem/MW-year is obtained for the first 10,000 years with uranium fuel. With recycled plutonium and uranium fuel (B 2) 0.33 manrem/MW-year is obtained.

Sensitivity Analysis of Some Coefficients in the GETOUT and BIOPATH Models

The GETOUT Model

The calculations made with the GETOUT model revealed that dispersion is an important parameter, especially with short leaching times. Dispersion means that if a water front starts passing through rock, part of it will arrive at a given point sooner than the main flow. Therefore, a short release from the reposi-

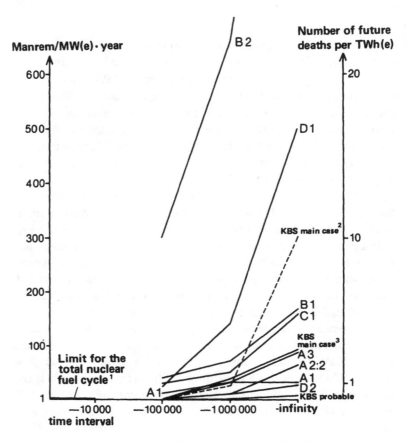

Fig. 14.3 Total global and cumulative dose commitments (in manrem per megawatt per year). Values are stated for total individual doses to world population during different time intervals and the first 10,000 years, 100,000 years, and so on. Scale on right side gives estimation of total deaths that might in future be caused by a repository of highly active waste. Note that these deaths are distributed over long periods. The recalculation factor from manrem to death is made in same way as in SSI's remiss answer (see also "Estimated Radiation Doses" above).

tory will show up in the recipient as an input over a long time, maybe several years, having a peak activity lower than that of the original release.

In the calculations dispersion is described by a diffusion coefficient. The basis for choosing that parameter has not been documented, and it is therefore not possible to review. Nevertheless, the importance of dispersion is shown in table 14.3 where different diffusion coefficients are used. The table shows that the activity inflow at short leach times differs by more than a factor of ten.

Grundfeldt[3] points out that the model is sensitive for the very short leaching times that have been used in one part of the calculations in the sensitivity analysis. Two aspects that, according to Grundfeldt, have not been regarded in the GETOUT calculations are the geographical extension of the repository and the correction that the model is one dimensional. These factors would imply that the widening of peaks in the GETOUT computations is underestimated.

All the calculations in this chapter have used the average value of dispersion through two zones—the buffer zone and the rock zone—that is, used it in the same way as KBS.

TABLE 14.3
IMPORTANCE TO THE INDIVIDUAL OF INFLOW DISPERSION

Leach time of the glass	Inflow to recipient of Cs-135, Ci/y	
	KBS values*	Changed values**
300,000 years	$7.4 \; 10^{-7}$	$7.8 \; 10^{-7}$
30,000 years	$2.3 \; 10^{-6}$	$7.8 \; 10^{-6}$
3,000 years	$2.4 \; 10^{-6}$	$6.4 \; 10^{-5}$

*The calculations are based on the use of an average value of the dispersion in the buffer zone and rock.

**Only the dispersion in rock has been calculated. Suggested by Bertil Grundfeldt for this sensitivity analysis and used in the calculations for Scandpower (SKI's remiss answer).

The difference between the two diffusion assumptions can also be illustrated with the calculated case (B 1) where the values have been chosen to cover the uncertain intervals according to the review. The calculations that have been made in the sensitivity analysis (the left column in table 14.3) give 100 mrem/year in the well alternative. If only the dispersion in rock is included (the right column) 230 mrem/year is obtained.

The BIOPATH Model

Calculations have been done where some of the more than 1,000 coefficients in the model have been altered. The concentration and distribution factors have been changed to the upper limit of the interval if there are different values reported in the literature (as in KBS-TR-40) and with a factor of ten otherwise. These factors describe the uptake of nuclides, for instance from soil to plants, from lake to fish, from plants to meat, and so forth (table 14.4).

In KBS main case this change means that the maximum individual dose is increased from 13 to about 30 mrem/year while increasing in the B 1 case from 100 to 270 mrem/year in the well alternative and in B 2 from about 1,000 to 2,700 mrem/year.

TABLE 14.4
DOSE INCREASES WHEN CONCENTRATION AND DISTRIBUTION
FACTORS INCREASE
(Increase Expressed in Percentage of Original Case)*

	Individual dose	Collective dose
Elements		
I-129	760	970
Cs-135	1,350	170
Ra-226	440	650
Ra-226, daughter to U 234	450	1,000
Np-237	260	1,000
Pu-239	270	110

*Estimated with an inflow of 10^{-6} Ci/year in the well case.

TABLE 14.5
Synthesis of Resulting Calculations

Case	Well Case		Lake Case		Maximum annual collective dose
	mrem/year	time*	mrem/year	time*	Manrem/year
A1	19 (Pu239)	1.4×10^4	8.9(Pu239)	1.4×10^4	42(Pu239)
A2:1	62(Tc99)	1.6×10^4	2.3(Tc99)	1.6×10^4	41(Tc99)
A2:2	32(Tc99)	1.7×10^4	1.4(Tc99)	1.7×10^4	22(Tc99)
A3	1.4(Np237)	1.6×10^5	0.37(Ra226)**	6.8×10^4	9.2(Ra226)**
B1	100(Pu239)	6×10^4	90(Pu239)	7.3×10^4	220(Pu239)
	110 (summed up)				
B2	1000(Pu239)	6×10^4	870(Pu239)	7.3×10^4	2100(Pu239)
C1	58(Np237)	2.1×10^4	49(Pu239)	7.3×10^4	120(Pu239)
C2	58(Np237)	1.1×10^5	5.5(Ra226)**	6.0×10^3	20(Ra226)**
			4.6(Pu239)	4.1×10^4	
D1	33(Np237)	1.2×10^5	14(Ra226)**	3.6×10^4	96(Ra226)**
D2	0.48(Np237)	1.5×10^5	0.24(Ra226)	2×10^5	3.8(Ra226)**
"KBS main case"	8.8(Np237)	1.1×10^5	0.7**		24(Ra226)**
"KBS prob"	0.14(Np237)	1.7×10^5	0.037(Ra226)***	6.8×10^4	0.93(Ra226)**

*The year with maximum dose
**Daughter to U234
***According to KBS

TABLE 14.6
SYNTHESIS OF RESULTING CALCULATIONS
(Total Collective Dose Commitments)

| Case | Manrem for the total repository | | | |
	The first 10,000 years	The first 100,000 years	The first 1,000,000 years	Infinity
A1	1.3×10^5	3.8×10^6	1.0×10^7	1.0×10^7
A2:1	0	2.8×10^5	3.5×10^6	1.9×10^7
A2:2	0	2.9×10^5	3.4×10^6	1.9×10^7
A3	2.0×10^3	3.5×10^5	9.3×10^6	2.6×10^7
B1	8×10^4	1.2×10^7	2.2×10^7	5.1×10^7
B2	1×10^5	8.9×10^7	2×10^8	6.7×10^8
C1	0	9.4×10^6	1.5×10^7	4.7×10^7
C2	0	0	6.0×10^6	9.1×10^6
D1	0	6.8×10^6	4.3×10^7	1.5×10^8
D2	0	1.2×10^5	3.5×10^6	8.0×10^6
KBS main case	1.7×10^4	2.0×10^6	1.1×10^7	2.8×10^7
KBS probable case	0	2.8×10^4	9.6×10^5	2.3×10^6

* Wastes corresponding to 300,000 MW(e)-a.

15

Has It Been Shown Where the Wastes Can Be Stored Finally?

The Stipulation Act requires that it be shown not only how wastes can be stored finally but also where. KBS conducted a series of drillings and testing programs at five sites to examine their suitability and concluded that three of the sites were, in fact, examples where a final repository might be located.

Explicit criteria for site selection have not been established anywhere in the world. KBS has not done so either, but implicitly the properties of a site are reflected by the parameter values used in the safety analysis. A site might be regarded as acceptable *if* its properties in all respects were equal to or better than these parameter values, *if* the models where these parameters are used represent actual conditions, and *if* the results obtained from these models are acceptable.

These ifs lead to some questions. The first if asks whether a site actually exists. Is there a large enough piece of rock to meet the parameter values specified, and how is this to be verified? Is the site, once located, accessible? Is it available, or are there conflecting claims, such as ore deposits in the immediate vicinity? The second if was discussed in conjunction with geology and the GETOUT model (chapters 13 and 14), and the third in the section on what is absolutely safe (chapter 8).

University of Uppsala writes in its review:

> The repository needs an area of about 1 km² in the horizontal plane. It is not possible to find a crackfree rock area of that size. The rock that will surround the final storage will therefore necessarily be penetrable by groundwater. The task is to choose an area where the frequency of cracks is small and where the cracks are relatively tight and show small continuity so that the crack system will give poor conductivity to the groundwater. The three areas on which the examinations have concentrated are chosen considering favorable crack density and stability and are situated far from the earthquake area in Sweden.

Winchester discusses the required size of a granite block to hold a 1 km² repository. This has not been discussed by KBS, and Winchester argues that a distance of up to 3 km in each direction from the repository might be desirable. The required size of a suitable block might therefore be 50–100 km².

Geological Survey of Canada writes in its review:

> Lack of an integrated broad overview of the fractures and fracture characteristics as they relate to permeability precludes any conclusions about the suitability of the study areas. Further, the assumptions required to outline flow rates and flow patterns are not realistic and do not correspond to the measured data.

> These factors indicate that the presented work should not, and indeed can not, be used to support an argument that a safe geologic repository has been defined.

The last statement is implicitly supported by several

Swedish reviewers. University of Uppsala concludes that "it seems likely that it will be possible to identify rock areas of sufficient magnitude with next to no water-conducting cracks."

Even if more extended investigations are motivated, they conclude that there is no basis for doubt that the waste can be stored in rock without dispersion of dangerous quantities of radioactive material.

SKI (Scandpower) judges that "it seems probable that areas and geological formations may be located in the Swedish bedrock where the properties specified by KBS are present."

SGU and PRAV take similar positions. Also the Norwegian Geological Survey says, "it is clear from our investigation that we think it will be possible to locate areas in Norway which meet the preferred criteria for final deposition in geological formations with crystalline bedrock."

However, Scandpower points out that it is not possible to demonstrate the suitability with only three boreholes in an area of 1 km^2. As discussed in the geology section, some reviewers also note that no drillings have been undertaken to depths below the level of the repository and they stress the need to do so.

The Swedish geologists consulted by the Nuclear Inspectorate have raised objections to two of the three areas proposed by KBS. The Finnsjö area is located close to an old iron ore area. High concentrations of chloride have been found in a nearby borehole. Krakemala is located in an area where valuable ores might be located in the future. For both areas, the boreholes have shown deep crushzones, casting doubt on whether a large enough area can be located. The third area, Karlshamn, is more promising, but since only one borehole has been drilled so far, it is too early to judge the suitability of the site. According to the inspectorate, the good experience from underground cavities in the area should be included in the assessment.

Aikin and Hare of Canada expect many acceptable sites to be present in Sweden and that most areas would have better rock than represented by a 400-year groundwater transport time. However, the only way to verify this is to identify a site, to bore, and to test.

Winchester, in his review for the Energy Commission, doubts that an acceptable site can be located:

> The question of finding a suitable location for the final waste repository is not answered directly by the KBS. Whereas the rock permeability results from a few bore holes presented for the Karlshamn, Finnsjö and Krakemala areas and for the Stripa mine indicate that sections of quite impervious granite or gneiss can be found in Sweden, the results of the geologic studies (reported in KBS technical reports 17 and 18, for example) stress the need to find a large enough block for a repository, and this may be situated within a system of faults. From our reading of the geologic evidence we find it by no means clear that a crystalline rock mass can be found with the required size and overall tightness which at the same time meets the other requirements of groundwater composition, temperature, freedom from disturbances from previous human activities and availability from a human standpoint.[1]

The information needed to judge a prospective site is discussed in U.S. Geological Survey circular 779, where the identification of a suitable site is also discussed:

> Confidence that a prospective site is likely to provide adequate containment derives from an understanding of structural and lithologic features apparent at scales from a few meters to tens of kilometers. Many critical features are subtle. For example, small faults, particularly strike-slip faults, can be extremely difficult to detect and to date. Fracture systems, which could act as short-circuiting conduits for ground water, might be revealed from regional structure features clearly evident only at the scale of a Landsat image, or perhaps from buried structural features indicated by the distribution of soil moisture. Furthermore, the necessary structural integrity and lithologic homogeneity cannot be assured simply through detailed surface geologic mapping and coring, particularly if the number of core holes is

limited by the risk to the potential repository caused by extensive local drilling.

Techniques for remotely exploring a volume of rock are in various stages of development and should be applied to potential waste repository sites as soon as practical. These methods are used from the surface, between boreholes and at the head of an excavation. They include high-resolution seismic and acoustic techniques, which can detect fine-scale structural and lithologic variations, and electric and electromagnetic methods, which are primarily sensitive to the distribution of water. Some of the most recently devised methods use shortpulse radar and continuous-wave interferometry; their use is limited to high-resistivity materials such as salt and perhaps dry crystalline rock.

The techniques cited provide a nondestructive way of characterizing the site in detail. Some of the borehole-to-borehole electrical techniques need to be further refined, but the critical need to map the original sub-surface structure, lithology, and ground-water regime warrants the aggressive utilization of such techniques where appropriate.[2]

The California Energy Commission finds, with respect to these considerations, that "the KBS project appears to have used fairly simple and probably inadequate techniques."

The Commission notes, as have others, the difficulty of establishing a site without destroying the very feature one seeks, particularly as the number of boreholes is limited by the risk of destroying the potential repository.

KTH writes:

To sum up it can be stated that the groundwater in rocks follows existing crack systems in the rocks. With present techniques it is not possible to determine completely safely in detail the crack system in an area, or the new cracks and crack systems that may be created during a long period of time. What is especially difficult is to determine, if a crack system is continuous and has communication with systems closer to the surface, so that a transport of nuclear waste to the surface may occur in a relatively short time.

Wynne-Edwards states:

Drilling and permeability tests on drill cores are unlikely, therefore, to reveal the true groundwater regime at the actual site, which may be dominated in its behavior by a few major fractures undetected until excavation. . . .

Theoretical studies based on rock properties are thus likely to be less reliable than a thorough examination and testing of the excavation itself. Given sound rock, however, and the absence of main fracture zones, the estimates of flow rates are reasonable.

At the Energy Commission hearing on June 8, Arne Wesslén (VIAK) used a metaphor: "Most geologists are convinced that there is copper ore for mining in Bergslagen. But nobody knows where. Prospecting has been carried out without any success for many years. So, even if we think that the necssary bedrock for a repository exists, that is not to say that we can locate it without very extensive prospecting."

Several reviewers are of the opinion that it is probable that a suitable site for a repository could be located. Others are not. There appears, however, to be a unanimous opinion that the existence of an acceptable site has not been proved.

PART V

Appendix

Abbreviations

AB Atomenergi Swedish agency for nuclear power development. The name has recently been changed to Studsvik Energiteknik AB.

AGNS Allied General Nuclear Services, company formed by Allied Chemical and General Atomic to construct and operate the Barnwell, South Carolina, U.S.A., reprocessing plants (BNFP).

AIB/VIAK Allmänna Ingenjörsbyrån AB and VIAK (Via et Aqua AB) Consulting firms (Consultants to the Swedish Nuclear Inspectorate during the KBS review)

AKA "Utredningen om radioaktivt avfall" (A several-year study of the general issues associated with radioactive waste management)

BNFP See AGNS

COGEMA

Compagnie Générale des Matières Nucléaires (French State company with which the Swedish utilities have a reprocessing contract)

CTH

Chalmers tekniska högskola (Chalmers Technical University, located in Gothenburg, Sweden).

EKA

Expert Group on Safety and Environment, Swedish-Energy Commission

EPA

Environmental Protection Agency (U.S.A.)

FOA

Försvarets Forskningsanstalt (National Defence Research Institute)

IAEA

International Atomic Energy Agency

ICRP

International Commission on Radiological Protection

INFCE

International Nuclear Fuel Cycle Evaluation

KBS

Kärnbränslesäkerhet (Nuclear Fuel Safety Project, the authors of the proposed Swedish method for management of vitrified liquid waste from reprocessing)

KTH

Kungl Tekniska Högskolan (Royal Institute of Technology, located in Stockholm, Sweden)

MFRP

Midwest Fuel Recovery Plant, the reprocessing plant at Morris, Illinois, U.S.A., built by General Electric Co. The plant is not in operation.

NEA — Nuclear Energy Agency, OECD's nuclear power agency.

NFS — Nuclear Fuel Services, company that operated a reprocessing plant at West Valley, New York, U.S.A. The plant is not in operation.

NPL — National Physical Laboratory (U.K.)

NRC — Nuclear Regulatory Commission (U.S.A.)

OECD — Organization for Economic Cooperation and Development, composed of several Western European countries and others outside of Western Europe. Parent organization for the NEA.

PRAV — Programrådet för radioaktivt avfall (Official organization having some responsibility for radioactive waste management in Sweden)

RINGHALS — A Swedish nuclear power station located at Värö, about 50 km south of Gothenburg. Owned and operated by the Swedish State Power Board, Vattenfall or Statens vattenfallsverk.

SCANDPOWER — Consulting firm in Norway which has assisted SKI in its review work. Scandpower coordinated the review.

SGI

Statens Geotekniska Institut (Swedish Geotechnical Institute)

SGU

Sveriges geologiska undersökning (The Swedish Geological Survey)

SKBF

Svensk Kärnbränsleförsörjning AB (Swedish company having responsibility for the fuel used in Swedish power reactors. The company is owned jointly by those Swedish electrical utilities that operate nuclear power stations.)

SKI

Statens kärnkraftinspektion (The Swedish Nuclear Inspectorate, the Swedish agency for nuclear power safety)

SSI

Statens strålskyddsinstitut (The Swedish National Institute of Radiation Protection, the Swedish agency for regulation of ionizing radiation)

STUDSVIK

AB Atomenergi's research laboratory located 80 km southwest of Stockholm)

SVAHN-COMMITTEE

A special committee that was appointed by the Swedish minister of energy in 1978 to coordinate an international review of the KBS-I report. The chairman of the committee was Hans Svahn, chief legal officer for the Ministry of Industry.

UNITED REPROCESSORS, GMBH

In 1971 Great Britain, France, and West Germany formed GMBH to coordinate their reprocessing development.

UNSCEAR   United Nations Scientific Committee on the Effects of Atomic Radiation

UP3-A   The reprocessing plant planned for construction in La Hague

USEPA   United States Environmental Protection Agency

USGS   United States Geological Survey

Reviewers

List of Swedish Reviewers (Participants in Remiss)

Försvarets forskningsanstalt (FOA)
Statens geotekniska institut
Sveriges meteorologiska och hydrologiska institut (SMHI)
Universitetet i Uppsala
Universitetet i Lund
Universitetet i Göteborg
Universitetet i Stockholm
Universitetet i Umeå
Universitetet i Linköping
Tekniska högskolan i Stockholm (KTH)
Chalmers tekniska högskola (CTH)
Högskolan i Luleå
Statens naturvårdsverk (SNV)
Statens strålskyddsinstitut (SSI)
Arbetarskyddsstyrelsen
Statens planverk
Sveriges geologiska undersökning (SGU)
Statens kärnkraftinspektion (SKI)

Styrelsen för teknisk utveckling (STU)
Programrådet för radioaktivt avfall (PRAV)
AB Atomenergi
Ingenjörsvetenskapsakademien (IVA)
Korrosionsinstitutet
Vetenskapsakademien
Jordens Vänner och Miljöförbundet

List of Foreign Reviewers

The review comments and addresses of the reviewers are published in DsI 1978:28.

General

Royal Society, England

Dr. T. English, Professor L. Lees, Jet Propulsion Laboratory, U.S.A.

IAEA, Austria

California Energy Resources Conservation and Development Commission, U.S.A.

Dr. A. M. Aikin, Professor F. K. Hare, Canada

Sir Brian Flowers, England

Encapsulation, Corrosion, and Creep

National Physical Laboratory, England	Glass
National Corrosion Service, England	Corrosion
Professor I. Oudar, Ecole Nationale Superieur de Chimie, France	Corrosion
National Engineering Laboratory, Scotland	Creep

Geology, Hydrogeology, Geochemistry, Retardation, Shafthole Plugging

Geological Survey of Canada, Canada
U.S. Geological Survey, U.S.A.

Generaldirektör K. Heier, Norwegian Geological Survey, Norway	Geology, hydrogeology, and geochemistry
Professor K. Rankama, University of Helsinki, Finland	Geology, hydrogeology, and geochemistry
Professor C. Fairhurst, University of Minnesota, U.S.A.	Crack formation and stability in tunnels and shafts
Professor H. R. Wynne-Edwards, University of British Columbia, Canada	Crack formation and stability in tunnels and shafts
Professor T. L. Brekke, University of California, U.S.A.	Stability in tunnels and shafts
Lawrence Livermore Laboratory, U.S.A.	Leaching of glass and retention of radionuclides
Professor G. R. Choppin, Florida State University, U.S.A.	Leaching of glass, chemistry, and transport of radionuclides
Professor Dr. Gerd Lüttig, Geological Surveys, Federal Republic of Germany	Shaft-hole plugging
Dames & Moore, England	

Dispersal of Radioactivity in the Biosphere

Dr. George Woodwell, Marine
Biological Laboratory, U.S.A.

National Radiological Protection
Board, England

Safety Analysis

Energy Incorporated, U.S.A.

Notes and References

All quotations in this book from the reviewers listed above are taken from "Report on Review through Foreign Expertise of the Report 'Handling of Spent Nuclear Fuel and Final Storage of Vitrified High Level Reprocessing Waste,' " Ministry of Industry, DsI 1978:28 (in English) and "Yttranden över statens vattenfallsverks ansökan enligt villkorslagen om tillstånd att tillföra reaktoranläggningen Ringhals 3 kärnbränsle," Industridepartementet DsI 1978:29 (in Swedish).

Several other Swedish government reports are included in these references. These reports may be obtained as follows:

DsI Reports

The *DsI* reports are a series published by the Swedish Ministry of Industry. They may be obtained from:
Industridepartementet
Fack
S-103 10 Stockholm

SOU Reports

The *SOU* reports are a series of official government publications. They are available from any Swedish bookseller. If there is difficulty in locating these reports, contact the Industridepartementet.

Prop and *NU*   These are reports of the Swedish Parliament. The *Prop* reports are materials submitted with draft bills at the time of their introduction. The *NU* reports are reports of a committee of the Parliament. They may be obtained from:
Riksdagen
Fack
S-100 12 Stockholm

4. The Legal Background

1. Proposition (of a Bill to the Swedish Parliament), *Prop* 53 (1976–77): 2–30.
2. Ibid., pp. 23–24.
3. Ibid., p. 24.

5. The Nuclear Fuel Cycle

1. A. Larsson, Å. Hultgren, J. Lind, "Management of Radioactive Waste and Plutonium in the Swedish Perspective" (International Conference on Nuclear Power and Its Fuel Cycle, Salzburg), 2–13 May 1977 (English original).
2. *Använt kärnbränsle och radioaktivt avfall*, AKA, *SOU* 1976:30 p. 88. This report has also been published in English as *Spent Nuclear Fuel and Radioactive Waste*, *SOU* (1976):32.
3. S. M. Keeny, Jr., et al., *Nuclear Power Issues and Choices* (Cambridge, Mass.: Ballinger Publishing, 1977) (often called the Ford/MITRE Study), pp. 249–250.
4. Ibid., p. 248.
5. Ibid., p. 267.

6.   H. C. Burkholder et al., *Incentives for Partitioning High-Level Waste*, BNWL-1927 (Richland, Washington: Battelle Pacific Northwest Laboratories, November 1975).

6.   How to Interpret "Has Shown"

1.   See, for example, "Status of Nuclear Fuel Reprocessing Spent Fuel Storage and High-Level Waste Disposal," draft report, California Energy Resources Conservation and Development Commission, January 11, 1978.

2.   T. Westermark, *DsI* (1978):17, appendix II (English original).

7.   What Constitutes "Highly Radioactive Waste . . ."

1.   *Prop* 53 (1976−77) 13.

2.   Ibid., p. 23.

3.   SSI remiss, p. 11.

4.   KBS-I, 1:14−16.

5.   SKI, pp. 30−31.

6.   SSI remiss p. 12.

7.   J. Rydberg and J. W. Winchester, "Disposal of High Active Nuclear Fuel Waste, A Critical Review of Nuclear Fuel Safety (KBS). Project on Final Disposal of Vitrified High Active Nuclear Fuel Waste," *DsI* 17 (1978):17, section 3.1.4 (English original).

8.   Ibid., II:18 (English original).

8.   What Is "Absolute Safety"

1.   *Näringsutskottets betänkande. NU* 23(1976−77) 15. (Parliament committee report on Stipulation Act.)

2.   Prop. 53(1976−77):240.

3. Statens stralskyddsinstitut, "*Begränsning av utsläpp av radioaktiva ämnen från kärnkraft-stationer*" (1977), p. 33. (This report is also published in English as "Limitation of Releases of Radioactive Substances from Nuclear Power Stations: The Swedish Regulations with an Introduction to Their Background and Purpose." Stockholm: The Swedish National Institute of Radiation Protection, 1977.)

4. Ibid., p. 34.

5. *Statens strålskyddsinstituts författningssamling*, SSI FS 2(1977):2.

6. SSI, "Begränsning . . .,", p. 24.

7. *Energi, Hälsa, Miljö, SOU* (1977):69 attachment 2, p. 101.

8. B. Persson, *DsI* (1978):24, report no 3.

9. Energy Commission, Expert Group on Safety and Environment (EKA), "Miljöeffekter och risker vid utnyttjande av energi," *DsI* (1978):27 p. 21:68.

10. J. Rydberg, *DsI* (1978):17, pp. IV:44—60 (English original).

11. Lars Norberg, Memorandum to EKA reference group 78-04-11 (English original).

12. T. Cochran, "Radioactive Waste Disposal Criteria," unpublished memorandum, 31 May 1978.

13. KBS 4: figs. 3-8, 3-9, pp. 28—29.

14. EKA *DsI* (1978):27, 2:21:69 and 21:70.

15. J. Rydberg, *DsI* (1978):17, sect. 4, fig. 13.

16. M. Hillert, cited in *DsI* (1978):17, IV:22 (English original).

13. Final Repository

1. L. Charles Hebel (chairman) et al., "Report to the American Physical Society by the Study Group on Nuclear Fuel Cycles and Waste Management" in *Reviews of Modern Physics* 50, no. 1, part 11, January 1978.

2. B. Biletti and R. Siever (cochairmen) et al., "The State of Geological Knowledge Regarding Potential Transport of High-Level Radioactive Waste from Deep Continental Repositories," Report of Ad Hoc Panel of Earth Scientists,

Office of Radiation Programs, U.S. Environmental Protection Agency, EPA/520/4-78-004, June 1978.

3. "Geologic Disposal of High-Level Radioactive Wastes in Earth-Science Perspectives," United States Department of the Interior, U.S. Geological Survey, Geological Survey Circular 779 (1978):8−9.

4. EPA, "Geological Knowledge," p. 44.

5. Ibid. pp. 44−45.

6. J. W. Winchester, Statement to Swedish Energy Commission Hearings, 8 June 1978.

7. KBS Technical Report Number 6, p. 63 (English original). Ulf Lindblom, "Groundwater Movements Around a Repository: Phase 1, State of the Art and Detailed Study Plan," Hagkonsult AB, Stockholm.

8. Personal communication: Ronny Bergman and Ulla Bergström (Studsvik Energiteknik AB).

9. A paragraph is here omitted from the original version as the discussion there was based on data from Studsvik Energiteknik AB, later withdrawn as being incorrect.

14. Sensitivity Analysis

1. Values for the high-level waste activity from the various transuranic nuclides are taken from GESMO: U.S. Nuclear Regulatory Commission, *Final Generic Environmental Statement on the Use of Recycled Plutonium in Mixed Oxide Fuel in Light Water Cooled Reactors*, NUREG-0002, August 1976.

2. Ibid.

3. B. Grundfeldt, Consultant to KBS, in Letters to the Ministry of Industry, 19 July 1978 and 21 July 1978.

15. Has It Been Shown . . .

1. J. Winchester, *DsI* (1978):17, p. IV−27 (English original).

2. USGS, "Geologic Disposal of High-Level Radioactive Wastes," p. 4.

Handling of Spent Nuclear Fuel and Final Storage of Vitrified High-Level Reprocessing Waste: Summary

(Reprint of the Summary Done by Nuclear Fuel Safety Project)

In April 1977 the Swedish Parliament passed a law that stipulates that new nuclear power units can not be put into operation unless the owner is able to show that the waste problem has been solved in a completely safe way. The task of investigating how radioactive waste from a nuclear power plant should be handled and stored was previously the responsibility of the National Council for Radioactive Waste Management (PRAV). This council was formed in November 1975 as the result of a proposal made by the Government Committee on Radioactive Waste (the AKA Committee).

In response to the government bill proposing the law, the power industry decided in December 1976 to give top priority to the investigation of the waste problem in order to meet the requirements of the law. Therefore, the Nuclear Fuel Safety Project (KBS) was organized. The first report from the KBS project entitled "Handling of Spent Nuclear Fuel and Final Storage of Vitrified High-Level Reprocessing Waste" was submitted in December 1977.

The Requirements of the Law Regarding Completely Safe Storage

The law stipulates that the owner of a reactor must show how and where completely safe storage can be provided for either the high-level reprocessing waste or the spent, unreprocessed nuclear fuel. "The storage facility must be arranged in such a way that the waste or the spent nuclear fuel is isolated as long a time as is required for the activity to diminish to a harmless level." "These requirements imply that measures should be taken which, during all phases of the handling of the spent nuclear fuel, can ensure that there will be no damage to the ecological system."

In the strictest meaning of the word, no human activity can be considered completely safe. The fact that such an interpretation of the wording of the law was not intended is evident from the formulation of the statements made by the government in support of the law indicating that the storage of waste shall fulfill "the requirements imposed from a radiation protection point of view and which are intended to provide protection against radiation damage." Questions regarding protection against radiation damage are regulated by the Radiation Protection Act. This means that the requirements imposed on the handling and storage of high-level waste are, in principle, the same as those that apply for other activities involving the handling of radioactive substances.

This interpretation is supported by the statements made by the Committee of Commerce and Industry in its review of the law, in which the Parliament also concurred. The Committee thus finds the expression "completely safe" to be warranted in view of the very high level of safety required, but considers that a "purely

Draconian interpretation of the safety requirement" is not intended. *Draconian* means "excessively severe, inhuman."

The Requirements of the Law Regarding the Scope of This Report

In the statements made by the government in support of the law it is said: "The descriptions to be submitted by the owner of the reactor shall include detailed and comprehensive information for the evaluation of safety. Consequently, overall plans and drawings will not suffice. Furthermore, it should be specifically stated in which form the waste or spent nuclear fuel is to be stored, how the storage is to be arranged, how the transportation of the spent nuclear fuel or of the waste will be carried out and whatever else may be required in order to ascertain whether the proposed final storage can be considered completely safe and possible to construct."

To fulfill these requirements, this report presents relatively detailed information on the design of facilities and the transportation systems that are part of the handling and storage chain. Certain parts of this information are relatively inessential for evaluating the safety of the waste storage, while others are vital. A detailed evaluation of the safety aspects of the proposed design is presented in a safety analysis. The handling and processing carried out abroad are also described, although more generally.

The Alternatives Given in the Law

The law requires a description of the handling and final storage of either the high-level reprocessing waste or the spent, unreprocessed nuclear fuel. This report deals with the first alternative. An application to the government to charge nuclear fuel to a new reactor based on this alternative must, in addition to this report, include an agreement that covers in a satisfactory manner the anticipated need for reprocessing of spent nuclear fuel. This aspect is, however, not dealt with in this report.

A report on the second alternative, that is, spent unprocessed fuel, is planned for publication during the first half of 1978.

Layout of the Report

This report has been divided into five volumes as follows:

I   General
II   Geology
III Facilities
IV Safety analysis
V   Foreign activities

In order to provide a basis for the report, KBS has carried out a great number of technical-scientific investigations and surveys. The results of these are published in KBS Technical Reports. Fifty-six volumes of these reports have been published so far (see volume I, appendix 3).

Volume I (General) can be read independently of the other volumes. It comprises mainly a summary of the more detailed reports presented in volumes II, III, and IV.

Chapter 3 in volume I is a summary of the proposed method for handling and storage of nuclear fuel and high-level waste from the nuclear power plant fuel pools up to and including final storage in Swedish bedrock.

Chapter 13 in volume I summarizes the more detailed presentation of the safety analysis in volume IV. This chapter summarizes the safety evaluations of the whole handling chain from a radiological point of view. The effects of radiation have been calculated for normal conditions and for accidents. Special emphasis has been placed on the long-term aspects of the final storage of high-level waste.

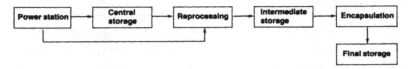

Final Stage of Nuclear Fuel Cycle

The handling chain for spent nuclear fuel and high-level reprocessing waste is illustrated in the above block diagram.

Nuclear power stations always have storage pools for spent nuclear fuel. They are needed so that the fuel can be discharged from the reactor and also to provide storage space for spent nuclear fuel before it is dispatched for reprocessing or for storage elsewhere.

Today the available reprocessing capacity is limited, and it is not clear to what extent spent nuclear fuel will be reprocessed. As a result, it is necessary to extend the storage capacity for spent nuclear fuel. For economic reasons and for the planning of the back end of the nuclear fuel cycle, the extended capacity should not be provided at the nuclear power stations. Instead, a central fuel storage facility should be constructed. This facility is needed regardless of whether the spent nuclear fuel is to be reprocessed or not before final storage. The fuel can be stored in this facility for about ten years.

As a rule, radioactive waste must be stored in the country where it is produced. The high-level reprocessing waste will be sent back to Sweden in vitrified form in 1990 at the earliest. The vitrified waste will be contained in stainless steel cylinders having a diameter of 40 cm and a height of 1.5 m. If all of the fuel is reprocessed, 9,000 cylinders will be obtained from thirteen reactors that have been in operation for thirty years.

The waste cylinders will be placed initially in an intermediate storage facility where they will remain for at least thirty years before being transferred to final storage. The cylinders will be kept in dry conditions in the intermediate storage facility, and radioactive substances cannot be released to the environment. During this storage period, the amount of heat generated by the waste will be reduced by half, thus simplifying final storage. Intermediate storage postpones the date when final storage must commence, thus providing more time to optimize the final storage method. A longer storage period than thirty years is entirely possible. Such a prolonged storage period is considered in France, for example. However, intermediate storage requires a certain amount of supervision, even though this supervision is very limited.

It is planned that the final storage, which will not have to go into operation until 2020 at the earliest, will be constructed in rock about 500 meters underground. The facility is designed in such a way that it can be sealed and ultimately abandoned. In the final storage, the waste will be exposed to the groundwater in the rock. After intermediate storage and before the waste cylinders are transferred to the final storage, they will therefore be encapsulated in a canister made of titanium and lead. These materials have good resistance to corrosion.

The siting of the facilities for the various handling stages may be arranged in different ways, in accordance with what is deemed to be practical.

Spent fuel has already been shipped abroad from Sweden for reprocessing. Similar transports will also be required between the various phases of the handling. The design and procurement of transport casks and vehicles thus form part of waste handling.

Geological Requirements for a Final Storage

Extensive investigations and tests have been carried out to determine the suitability of Swedish bedrock for final storage. In this connection, interest has been concentrated on precambrian crystalline rocks. In other countries, studies have been made of storage in salt, shale, and clay depending upon the natural prerequisites of each country.

Field investigations have been carried out at five sites, three of which have been selected for more detailed studies. A number of holes have been drilled to a depth of 500 meters. It should be emphasized that the objective of this work was not to find a site now to be proposed for final storage. The purpose was to show that suitable bedrock is available within Sweden for such a facility.

The factors that will determine the suitability of a rock formation for final storage are its permeability and strength, the composition of the groundwater and its flow pattern, and the delaying effects on radioactive substances when groundwater passes through cracks in the rock. Of special interest also is the risk of rock movements that could affect the pattern of groundwater flow or damage the encapsulated waste.

Assessing these factors, a depth of about 500 meters is considered to be suitable. At this depth, the bedrock contains fewer cracks and has lower water permeability than closer to the surface. This depth also gives satisfactory protection against acts of war and such extreme events as meteorite impacts and the effects of a future ice age.

The investigations and surveys carried out have shown that the three sites selected offer satisfactory conditions for final storage. At these sites, the bedrock consists of Sweden's most common types of rock: granite, gneiss, and gneissified granodiorite. Consequently, it is reasonable to expect that rock formations with equivalent conditions are also available at many other places within Sweden.

Safety of the Handling Chain

The extensive safety analysis carried out has shown that the release of radioactive substances that could occur in connection with normal operation or with an accident in the different stages of the handling chain within Sweden would be insignificant in comparison with corresponding conditions at a nuclear power station. This is because the vitrified waste has a low temperature and is encapsulated without overpressure. Consequently a sudden and extensive release of radioactivity cannot occur. The safety of the steps of the handling chain, which will be carried out abroad (reprocessing and vitrification), will be evaluated by government authorities in the country concerned and are dealt with in a more superficial manner in this report.

Radioactive substances from a final storage can only be released by the groundwater. The final storage must be arranged in such a way that such a release cannot damage the ecological system. It is then important to remember that the activity of the radioactive substances in the waste diminishes very slowly. The final storage is therefore arranged so that the migration of these substances is either prevented or delayed for a long time, thus ensuring that the concentration of radioactive substances that may reach the biosphere will be harmless. For this reason, the design of the final storage provides for a number of successive barriers.

For any release of radioactive substances in the waste to the environment, the groundwater must first penetrate both the canister made of titanium and lead and the stainless steel container.

These materials have excellent resistance to corrosion. The waste cylinders will be placed in holes drilled into good quality rock and surrounded by a buffer material consisting of quartz sand and bentonite. Since the buffer material has a low permeability, only very small amounts of water will be able to affect the encapsulated waste.

In the event of the penetration of the canister and the stainless steel container, the groundwater can affect the vitrified waste. However, the glass has a very low leaching rate under the conditions that prevail in the final storage.

The low flow rate of the groundwater, the long distance that the water must cover to reach the biosphere and the chemical processes in the crack system in the rock and in the buffer material provide effective barriers that prevent and delay the migration of the radioactive substances. Moreover, dilution in huge volumes of groundwater will take place before entry into the biosphere.

The safety of the final storage of high-level waste is dominating the safety issue. The safety analysis is based, in each phase that entails uncertainty, on assumptions and data that provide a reassuring margin of safety. Possible routes for the migration of radioactivity to the biosphere have been studied in the safety analysis, and the group of people who can be exposed to the highest level of radiation has been identified (the critical group). The critical group consists of persons taking their drinking water from a deep well drilled in the vicinity of the final storage. Under unfavorable circumstances this group can be exposed to a maximum radiation (individual dose) of 13 millirems per year in addition to natural background radiation.

This maximum additional dose of 13 millirems per year will not occur until after about 200,000 years. This long delay is caused by the retainment in the buffer material and the rock of the radioactive substances providing the highest additional dose. Radioactive substances that are not delayed relative to the flow of water in the bedrock could come into contact with the biosphere after only some hundreds of years. However, the additional dose attributable to these substances is very much lower than the value given above.

An individual dose of 13 millirems is considerably lower than the dose recommended by the International Commission on Radiological Protection (ICRP) as the upper limit for permissible additional doses for individuals, namely 500 millirems per year. This limit is intended to protect individuals against delayed radiation effects such as cancer and genetic effects.

Governmental authorities impose lower limits for the operation of nuclear power plants. In Sweden, operational restrictions can be imposed and other measures taken if the additional dose tends to exceed 50 millirems per year for people living near the power plant.

In order to reduce radiation exposure as much as reasonably possible, the Swedish Radiation Protection Institute requires that nuclear power plants be designed and constructed so that the expected additional dose for the critical group living in the vicinity of the plant is less than 10 millirems per year.

As mentioned above,the assumptions and data used in the safety analysis were selected with safety margins. It is considered probable that the dosage will be approximately 1/100th of the maximum value of 13 millirems per year given above. One reason for this is that the very low rate of water flow in the bedrock is not sufficient to break through the encapsulation or leach the vitrified waste at the rates assumed in the safety analysis presented in this report. However, verification of this lower value would require additional investigations not yet completed.

The following bar chart shows the dose rates mentioned above. It also indicates the dose rates from natural radiation in Sweden. As appears from the bar chart local variations in natural radiation are considerably greater than the maximum contribution from final storage of high-level waste obtained from thirteen reactors that have been in operation thirty years. The bar chart also shows that the doses obtained from radium in natural drinking water in Sweden often lie considerably above the level reported for final storage.

Moreover, the safety analysis shows that radiation doses for large population groups attributable to final storage will be virtually insignificant and that the long-term effects on health will be negligible.

The design of the back end of the nuclear fuel cycle presented in this report thus fulfills the requirements set forth in the law for a completely safe final storage of the high-level reprocessing waste.

Stockholm, November 1977
Nuclear Fuel Safety Project (KBS)

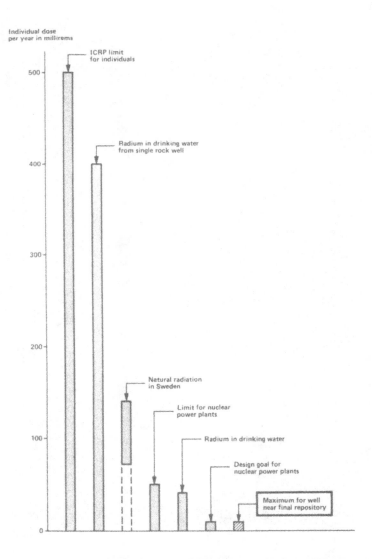

Individual dose
per year in millirems

500 —

ICRP limit
for individuals

400 —

Radium in drinking water
from single rock well

300 —

200 —

Natural radiation
in Sweden

Limit for nuclear
power plants

100 —

Radium in drinking water

Design goal for
nuclear power plants

Maximum for well
near final repository

0 —

Bar graph showing the calculated maximum annual radiation doses which the final repository can give to a nearby resident and the annual dose to man from some natural radiation sources plus some established dose limits. The dose from drinking water comes from radium-226.

Index

Absolute safety: criteria for, 67–68; and dose level, 70; interpretations of, 30, 37, 67–83; and KBS-I method, 15, 22–24, 39–41, 78–83; Nuclear Stipulation Act requirements for, 20, 30; and radiation exposure standards, 76; and sensitivity analysis, 15

AKA investigation, 64, 65

Americium, 72

Antinuclear position, 4–11 passim

Atomic energy program. *See* Nuclear power program

Barriers between radioactive elements and biosphere, 28, 79, 94–124; mechanisms acting upon, 105–124. *See also* Encapsulation cylinders; Encapsulation materials; Rock, as final repository

Bentonite, for plugging tunnels and shafts, 116–117

BIOPATH model, 79, 124–130; and dilution effects, 117; parameters included in, 137; sensitivity analysis of some coefficients in, 146, 152–154

Breeder reactors: introduction of, 4; plutonium as fuel in, 4, 23–24, 49–50

California Energy Commission: on leaching rates, 97; on Nuclear Fuel Safety (KBS) Project approach to radioactive waste disposal, 81–82; on transportation risks, 88

California Energy Resources Conservation, and California nuclear laws, 58–59

Carbon-14: released into atmosphere, 53; removal of, 134

Center Party, 5, 6, 9

Centrallager (central spent fuel storage pool), 21, 84–85

Central spent fuel storage pool. *See* Centrallager

Designer: Graphics Two
Compositor: Trend Western Corp.

Text: 10/13 Baskerville
Display: Helvetica Extra Bold